WATER RESOURCE PLANNING, DEVELOPMENT AND MANAGEMENT

GROUNDWATER QUALITY

ASSESSMENT AND ENVIRONMENTAL IMPACT

WATER RESOURCE PLANNING, DEVELOPMENT AND MANAGEMENT

Additional books and e-books in this series can be found on Nova's website under the Series tab.

WATER RESOURCE PLANNING, DEVELOPMENT AND MANAGEMENT

GROUNDWATER QUALITY

ASSESSMENT AND ENVIRONMENTAL IMPACT

RAFAEL M. VICK
EDITOR

Copyright © 2020 by Nova Science Publishers, Inc.

All rights reserved. No part of this book may be reproduced, stored in a retrieval system or transmitted in any form or by any means: electronic, electrostatic, magnetic, tape, mechanical photocopying, recording or otherwise without the written permission of the Publisher.

We have partnered with Copyright Clearance Center to make it easy for you to obtain permissions to reuse content from this publication. Simply navigate to this publication's page on Nova's website and locate the "Get Permission" button below the title description. This button is linked directly to the title's permission page on copyright.com. Alternatively, you can visit copyright.com and search by title, ISBN, or ISSN.

For further questions about using the service on copyright.com, please contact:
Copyright Clearance Center
Phone: +1-(978) 750-8400   Fax: +1-(978) 750-4470   E-mail: info@copyright.com

NOTICE TO THE READER

The Publisher has taken reasonable care in the preparation of this book, but makes no expressed or implied warranty of any kind and assumes no responsibility for any errors or omissions. No liability is assumed for incidental or consequential damages in connection with or arising out of information contained in this book. The Publisher shall not be liable for any special, consequential, or exemplary damages resulting, in whole or in part, from the readers' use of, or reliance upon, this material. Any parts of this book based on government reports are so indicated and copyright is claimed for those parts to the extent applicable to compilations of such works.

Independent verification should be sought for any data, advice or recommendations contained in this book. In addition, no responsibility is assumed by the Publisher for any injury and/or damage to persons or property arising from any methods, products, instructions, ideas or otherwise contained in this publication.

This publication is designed to provide accurate and authoritative information with regard to the subject matter covered herein. It is sold with the clear understanding that the Publisher is not engaged in rendering legal or any other professional services. If legal or any other expert assistance is required, the services of a competent person should be sought. FROM A DECLARATION OF PARTICIPANTS JOINTLY ADOPTED BY A COMMITTEE OF THE AMERICAN BAR ASSOCIATION AND A COMMITTEE OF PUBLISHERS.

Additional color graphics may be available in the e-book version of this book.

Library of Congress Cataloging-in-Publication Data

ISBN: 978-1-53618-807-3

Published by Nova Science Publishers, Inc. † New York

CONTENTS

Preface		vii
Chapter 1	Arsenic in Groundwater: Occurrences, Implications and Mitigation Strategies *Anadi Gayen*	1
Chapter 2	Groundwater Quality in the Megacity Dhaka, Bangladesh: Assessment and Environmental Impact *Shama E. Haque*	53
Chapter 3	From Rainwater to Groundwater Chemistry. Case Study: Mt. Cameroon Area *Kengue N. J. Dirane, Andrew A. Ako, Fozao K. Folepei, Bertil Nlend and Gloria E. T. Eyong*	79
Chapter 4	High Exposure Dose of Fluoride Owing to Risk of Fluorosis in Inhabitants of Sareni Block Located at the Ganga River Basin Uttar Pradesh (India) *Pokhraj Sahu, Pramod Kumar Singh, Prashant Singh, Vinay Kumar and Pramod Kumar*	107
Index		129

PREFACE

Groundwater Quality: Assessment and Environmental Impact first discusses arsenic contamination in groundwater, which has emerged as a major health hazard in India.

The authors review a generalized scenario of groundwater in the Greater Dhaka Area, focusing on the deterioration of groundwater quality over the years and its impact on the environment.

Following this, the chemical composition of rainwater and groundwater from the Mount Cameroon area in May-July 2017 is analyzed, and the impact of water-rock interactions on groundwater chemistry is assessed.

The concluding study aims to assess the extent of exposure to fluoride in inhabitants of Raebareli district in Uttar Pradesh, India, generating baseline data about the fluoride-contaminated area.

Chapter 1 - Arsenic is a naturally occurring contaminant that has known adverse human health effects. Drinking water is one of the most important pathways of exposure to arsenic in the human population and groundwater as a drinking source is thought to be responsible for the majority of the world's chronic arsenic-related health problems. Arsenic contamination in groundwater has emerged as a major quality problem and health hazard in states of the country. Nearly 40 million Indians are residing within the risk zone of arsenic contamination in groundwater.

Elevated arsenic content in groundwater is one of the most serious concerns particularly in the Ganga River Basin, where a study by Central Ground Water Board (CGWB) has shown that in the 17 States of India are arsenic affected. West Bengal is the most affected states in the India. The first case of arsenic-contaminated groundwater from the Bengal basin was reported in 1978 in the state of West Bengal. As in other parts of the country, groundwater in shallow aquifers is the major source of drinking water, wherein the maximum 300 ppb arsenic content has been reported. Not only for drinking, but groundwater from the shallow aquifer system is also responsible for recurring sources for most of the irrigated agriculture in this region. Researches reveal that arsenic is entering into the food chain, which is of great concern, especially for the rural population. Naturally occurring arsenic (As) in Holocene aquifers in the Indian sub-continent has undermined the long success of supplying the population with safe drinking water. Arsenic is mobilized in reducing environments through the reductive dissolution of Fe(III)-oxyhydroxides. Several studies have shown that many of the tested arsenic mitigation options have not been well accepted by the people. Instead, local drillers target presumed safe groundwater based on the color of the sediments. The overall objective of the study has thus been focussed on assessing the potential for local drillers to target arsenic safe groundwater. The initiative has to be undertaken immediately to address the arsenic problems in groundwater and site-specific measures need to be adopted on a priority basis for providing safe arsenic-free water to the people residing in the vulnerable areas of arsenic severity. A roadmap needs to be designed for addressing the arsenic issues and challenges for achieving a practicable sustainable solution and to develop a model to implement the arsenic mitigation steps through a well-coordinated approach.

Chapter 2 - Dhaka, the capital of Bangladesh, with a population of over 18 million is one of the fastest growing megacities of the world. The city faces many water resource management challenges that are common to other megacities of the 21st century. Due to rapid urbanization along with increased industrialization, the city struggles to provide sufficient water for its residents. Numerous surface water bodies in and around the city have been subjected to increased contamination and pollution due to indiscriminate disposal of untreated waste by municipal, industrial, commercial and agricultural sources. As such, in an effort to meet the hydrologic needs of its inhabitants, groundwater is being extracted from the Dupi Tila aquifer, which underlies the city. However, the current rate of groundwater withdrawal is beyond sustainable yields as indicated by a rapidly falling water table. In the past five decades, groundwater pumping in Dhaka has caused groundwater levels to drop more than 61 m and the average declination of static water level is approximately 3 m/year. Abstraction of groundwater at a faster rate than it can be recharged is likely to have negative impacts on the natural environment, including *permanently reduced aquifer storage capacity,* land subsidence, reduction of water in surface water bodies along with deterioration of water quality. This chapter aims to discuss a generalized scenario of groundwater in Greater Dhaka Area focusing on deterioration of groundwater quality over the years and its impact on the environment.

Chapter 3 - Rainfall may contribute to surface and groundwater sources. In this study rainwater and groundwater sampling was done and analyzed for chemical composition in the Mt. Cameroon area for the period May-July 2017. Water samples were investigated for their physico-chemical characteristics and assess the impact of water-rock interactions on groundwater chemistry in the Mt. Cameroon area. The wide ranges of EC values (3.1-367µS/cm) and total dissolved solids (1.74-108.03mg/L) revealed the heterogeneous distribution of hydrochemical processes within the groundwater of the area. The relative abundance of major cations and anions in the analyzed water (mg/L) is $Ca^{2+}>Na^+>Mg^{2+}>K^+>NH_4^+$ and $HCO_3^->Cl^->SO_4^{2-}>NO_3^-$, respectively. All the groundwater samples were soft, with total hardness values (2.54-136.65 mg/L) below the maximum permissible limits of the World Health Organization (WHO) guideline. A Piper diagram classified the water types into Ca-HCO$_3$ type (67%), Na-Cl type (17%) and the rest from the Na-Ca-HCO$_3$ and Ca-Mg-Cl water types; indicating recharge, seawater intrusion, and mixed respectively. Alkaline earth metal contents dominated those of alkali metals in most of the samples. Based on ion geochemistry, water-rock interactions, mixing, ion exchange and anthropogenic activity are the dominant hydrogeochemical processes. The weathering type corresponds to bisiallitization indicating the genesis of smectites where kaolinite and montmorillonite are the main minerals formed. A reasonable conclusion is that the groundwater chemistry is dominated by them and rainwater of local origin.

Chapter 4 - In India, during 1991, an estimate of 66 million was at risk of fluorosis which reached up to 120 million in 2018, it was an alarming situation and it seemed necessary to take some vital steps in order to address the fatal and incurable problem in the country. A study was conducted to assess the extent of exposure dose of fluoride in inhabitants of Sareni block of Raebareli district, Uttar Pradesh, where in, twenty-five groundwater samples were collected from different places in different direction (center, north, south, east and west zone) of Sareni block. The mean concentration of fluoride was found to be 1.19, 1.27, 1.23, 1.52, and 1.19 mg/L in center, north, south, east and west zones respectively, which is higher than Indian drinking water standard. The estimated dose of exposure of fluoride in the drinking water, as per the investigation in the drinking, were ~8, ~4 and ~ 2 times higher than the intake dose of fluoride recommended by Agency for Toxic Substances and Disease Registry for infants, children, and adults respectively. The results indicated that children were highly closed to the risk of incurable fluorosis as compare to adults in the study area because of high exposure and absorption efficiency of fluoride. Water quality assessment is one of the most important steps of water management practice for it is safe in drinking water uses or other specific purpose. This study will generates baseline data about the fluoride contaminated area of Sareni block.

In: Groundwater Quality
Editor: Rafael M. Vick

ISBN: 978-1-53618-807-3
© 2020 Nova Science Publishers, Inc.

Chapter 1

ARSENIC IN GROUNDWATER: OCCURRENCES, IMPLICATIONS AND MITIGATION STRATEGIES

Anadi Gayen, PhD[*]

Senior Hydrogeologist, Rajiv Gandhi National Ground Water Training and Research Institute, Central Ground Water Board
Department of Water Resources, River Development, and Ganga Rejuvenation, Ministry of Jal Shakti, Government of India
Naya Raipur, Chhattisgarh, India

ABSTRACT

Arsenic is a naturally occurring contaminant that has known adverse human health effects. Drinking water is one of the most important pathways of exposure to arsenic in the human population and groundwater as a drinking source is thought to be responsible for the majority of the world's chronic arsenic-related health problems. Arsenic contamination in groundwater has emerged as a major quality problem and health hazard in states of the country. Nearly 40 million Indians are

---

[*] Corresponding Author's E-mail: anadigayen1968@gmail.com.

residing within the risk zone of arsenic contamination in groundwater. Elevated arsenic content in groundwater is one of the most serious concerns particularly in the Ganga River Basin, where a study by Central Ground Water Board (CGWB) has shown that in the 17 States of India are arsenic affected. West Bengal is the most affected states in the India. The first case of arsenic-contaminated groundwater from the Bengal basin was reported in 1978 in the state of West Bengal. As in other parts of the country, groundwater in shallow aquifers is the major source of drinking water, wherein the maximum 300 ppb arsenic content has been reported. Not only for drinking, but groundwater from the shallow aquifer system is also responsible for recurring sources for most of the irrigated agriculture in this region. Researches reveal that arsenic is entering into the food chain, which is of great concern, especially for the rural population. Naturally occurring arsenic (As) in Holocene aquifers in the Indian sub-continent has undermined the long success of supplying the population with safe drinking water. Arsenic is mobilized in reducing environments through the reductive dissolution of Fe(III)-oxyhydroxides. Several studies have shown that many of the tested arsenic mitigation options have not been well accepted by the people. Instead, local drillers target presumed safe groundwater based on the color of the sediments. The overall objective of the study has thus been focussed on assessing the potential for local drillers to target arsenic safe groundwater. The initiative has to be undertaken immediately to address the arsenic problems in groundwater and site-specific measures need to be adopted on a priority basis for providing safe arsenic-free water to the people residing in the vulnerable areas of arsenic severity. A roadmap needs to be designed for addressing the arsenic issues and challenges for achieving a practicable sustainable solution and to develop a model to implement the arsenic mitigation steps through a well-coordinated approach.

Keywords: natural, contaminant, Holocene, aquifer, mitigation, sustainable

INTRODUCTION

Naturally, arsenic occurs in rocks, soil, water, air, plants, and animals. Arsenic enters into the physiological system through the source of drinking water and in turn causes drastic health hazards even in the form of cancer. Arsenic toxicity in groundwater is a global problem. Arsenic contamination in groundwater in the Indus-Ganga- Brahmaputra (I-G-B) fluvial plains in India and Padma-Meghna(P-M) fluvial plains in Bangladesh and its consequential effect on human health have been reported as one of the world's biggest natural groundwater disaster to the human population. As reported in India, the states like West Bengal, Jharkhand, Bihar, Uttar Pradesh situated in the flood plain of the Ganga River; Assam and Manipur states located in the flood plain of the Brahmaputra and Imphal rivers; and the consolidated formation in the Rajnandgaon district in Chhattisgarh state have been affected by Arsenic contamination in groundwater beyond the permissible limit of WHO as 10ppb. People residing in these affected states are severely exposed to drinking arsenic-contaminated water from the hand pump fitted tube wells. Every fresh survey reports more arsenic affected areas and people in the prey of arsenic-related diseases, and the arsenic-related issues and challenges are becoming much more complex on account of multiple ambiguous reasons. The flood plains/alluvial zones are representing the Holocene aquifers of recent geological age and have originated from the Himalayan mountains. Arsenic contamination in groundwater is responsible for multiple consequential perspectives like ingestion through the food chain, arsenicosis, cancer, social implications, and socio-economic problems in the affected regions. Over-extraction of groundwater is one of the key factors aggravating the arsenic challenges. The researches reveal that the occurrences of arsenic are sporadic in nature and geogenic in origin. The arsenic releasing mechanism involves the liberation of arsenic from soil under the conducive condition to arsenic solid phase followed by a liquid phase in groundwater. The irrigation return flow carrying fertilizers/pesticides/ herbicides/insecticides bearing water from the agricultural field enriched with arsenic is in turn percolating to sub-surface aquifer system and are causing a severity of the arsenic scenario further.

There are some hypotheses about the source of Arsenic and probable reasons for occurrence in groundwater. Arsenic is mobilized in reducing environments through the reductive dissolution of Fe (III)-oxyhydroxides. At present, the arsenic-related problems are being addressed partially in an inadequate manner, which requires strengthening by proper scientific support. Many researchers have undertaken innumerable R&D studies intending to make out the gravity of the arsenic menace, occurrence and distributions, health impacts, and related social aspects. They have come up with several finding include shortfalls to success stories. At present arsenic affected people are in dire need of the systematic implementation/ extrapolation of the successful studies in the areas of arsenic epidemics. Hence, more numbers of R&D studies are to be carried out to address the arsenic problems and remedial techniques, so that a comprehensive model can be designed in consideration with the area-specific occurrence and extension of arsenic in groundwater.

Table 1. Mendeleev's Periodic Table

Source: Wikipedia.

Global Scenario

High Arsenic concentrations in groundwater have been reported recently from the countries like USA, China, Chile, Columbia, Bangladesh,

Taiwan, Mexico, Argentina, Poland, Brazil, Zimbabwe, Canada, Hungary, Japan, and India. Among 21 countries in different parts of the world affected by arsenic contamination in groundwater, the largest population at risk is reported in Bangladesh followed by West Bengal in India (Mahan and Pittman, 2006). The occurrence of Arsenic in groundwater is not unique and as such Arsenic in groundwater has also been reported from different parts of the USA – in Alaska, Arizona, California, Idaho, Nevada, Oregon, and Washington. The concentration of Arsenic in the groundwater sample of Fallon New Naval Air Station (NAS) was found to vary from 0.08 to 0.116 ppm (Chadha and Sinha Ray, 1999). It was reported in the year 1960 that a large number of children were affected by Arsenic contamination in drinking water (0.08 ppm) in Antofagasta, Chile through leakage of arsenical water from mining operations into spring water. Arsenic contamination of streams and wells has been reported from the Obuasi gold mine area of Ghana. Endemic contamination of fresh water supply through leaching of Arsenic waste from mining operations has also been reported from Argentina. Arsenic is accumulated in the soils of extensive areas of Andes Mountains between Argentina and Chile and subsequently, it was found that the natural water sources have also been contaminated.

The occurrence of high arsenic has also been reported from the south-east part of Hungary, Baikalia province of Russia, and also from some parts of Newzealand and Cambodia. In Taiwan, a high concentration of Arsenic (0.53-1.19 ppm) in groundwater was reported in 1967 and a sizeable population was affected by arsenic dermatosis. In neighboring countries of Bangladesh groundwater in 59 districts out of 64 districts was reported with arsenic content varying from 0.01 ppm to above 0.05 ppm, where more than 51 million populations were affected. In Ronibook, Thailand, the Tin mine operations have contaminated well water with arsenic. This is due to leakage of arsenic caused by sulfuric acid, used for ore dressing being dumped into rivers. In Guizhou Province, China, Arsenic pollution has resulted from using coal with high concentrations of arsenic over a long period. In Juroku and Matsuo, Japan, highly poisonous white arsenic was produced due to the use of primitive calumniating furnaces for nearly half a century, causing ultimately pollution of air, soil, and river water. Soon after the residents along the south-west coast of Taiwan started using artesian wells for drinking in 1920, chronic arsenic poisoning, and known locally as Black Foot Disease (BFD) began. In Mindanao Island, the Philippines, arsenic was found in well and river water immediately after drilling in connection with a geothermal power plant on Mt. Apo in January 1992. The global scenario on arsenic distribution in groundwater is shown in Figure 1.

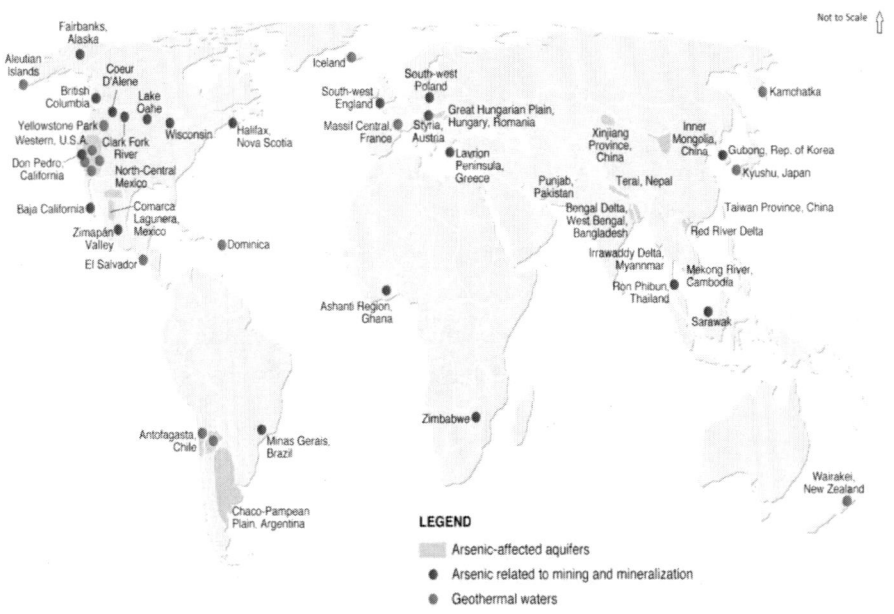

Sources: Bhatt, 2012.

Figure 1. Arsenic affected aquifers around the world

Groundwater environments governing mobilization of geogenic As can be broadly categorized in three groups (Smedley and Kinniburgh 2002; Sracek et al., 2001, 2004a): i) strongly reducing aquifers, ii) high alkaline- and pH in mostly oxidizing aquifers, and iii) aquifers containing elevated amounts of arsenopyrite and other sulfides (CGWB). Bioremediation method was used over 100 years ago with the opening of the first biological sewage treatment plant in Sussex, UK, in 1891 (NABIR Primer, 2003). Akhtar et al., (2013) gives an overview of Arsenic and lead contaminants in soil and also the mechanism of removal of these toxic metals from the contaminated sources by the potential application of plants and microbes. Electro-Chemical Arsenic Remediation (ECAR) is compatible with community-scale water treatment for rural Bangladesh. Addy et al., (2010), demonstrate the ability of ECAR to reduce Arsenic levels from >500 ppb to less than 10 ppb in synthetic and real Bangladesh groundwater samples and examine the influence of several operating parameters on Arsenic removal effectiveness.

Ahmad (2013) has been reviewed the disposal of water treatment wastes containing Arsenic, with a particular emphasis on stabilization/solidification (S/S) technologies, which are currently used to treat industrial wastes containing with. Water treatment wastes containing Arsenic are often mixed with cow dung in Bangladesh and India because the micro-organisms present in cow dung reduce soluble arsenic species to gaseous Arsine (AsH_3), which is released into the atmosphere (Ahmad, 2013). Incorporation of Arsenic sludge into construction materials is common in urban areas of Bangladesh and India (Ahmad, 2013). Thus the typical products include are cement blocks and cement plinths for latrines. Southern California Water Company (SCWC) found the GFH (Granulated Ferric Hydroxide) media was the most successful media at removing Arsenic and in meeting other performance criteria important to the system. Each of the Activated Alumina (AA) vessels experienced an Arsenic breakthrough in the effluent well before the GFH vessel (U.S. EPA, 2003). As has also been reported from the Great Alluvial Basin (GAB) in Hungary, Romania and Slovenia (Varsanayi et al., 1991, 2006, Lindberg et al., 2005, Petrusivski et al., 2007). Arsenic is also commonly encountered in oxidizing aquifers with high alkalinity and pH in the Chaco-Pampean region of Argentina affecting at least 1.2 million people (about 3% of the total population) are exposed to elevated As concentrations mobilized primarily from volcanic ash, interbedded or dispersed within sediments (Bundschuh et al., 2004, Bhattacharya et al., 2006b, Nicolli et al., 2010, 2012).

The Scottsdale Municipal Water system was also conducting additional studies using Granulated Ferric Hydroxide (GFH) and was indicated that GFH would be a promising treatment method, while Activated Alumina (the cheapest option) might not be, due to the high pH and silica levels in the raw water. Whereas both AA and GFH appear to be the best treatment options for Tucson water, which primarily depended on groundwater that is affected by naturally occurring Arsenic (U.S. EPA, 2003). Ali et al., (2003) conducted Toxicity Characteristic Leaching Procedure (TCLP) and column leaching tests (CLT) on waste materials taken from several removal units used in Bangladesh, and found that in general, leaching of As from the wastes is not significant and that none of the waste samples were hazardous as defined by United States Environmental Protection Agence (USEPA).

Indian Sub-Continent Scenario

Before 2000, there were five major incidents of arsenic contamination in groundwater in Asian countries: Bangladesh, West Bengal, India, and sites in China. Between 2000 and 2005, arsenic-related groundwater problems have emerged in different Asian countries in the Indian sub-continent (Figure 2), including new sites in China, Mongolia, Nepal, Cambodia, Myanmar, Afghanistan, DPR Korea, and Pakistan (Mukherjee et al., 2006). There are reports of arsenic contamination from the Kurdistan province of Western Iran and Viet Nam, where several million people may have a considerable risk of chronic arsenic poisoning. From a human health perspective, high As aquifers related to strongly reducing conditions pose most serious problems because of its wide geographical coverage especially in countries like Bangladesh, India, Pakistan, Vietnam, and Cambodia (Bhattacharya et al., 1997, 2001, 2002b, Ravenscroft et al., 2001, 2005, Smedley and Kinniburgh 2002, 2006a, Nickson et al., 2005, Berg et al., 2007). The arsenic scenario in groundwater in the Indian sub-continent is shown in Figure 2. In India, seven states namely- West Bengal, Jharkhand, Bihar, Uttar Pradesh in the flood plain of the Ganga River;

Assam and Manipur in the flood plain of the Brahmaputra and Imphal rivers, and Rajnandgaon village in Chhattisgarh state have so far been reported affected by Arsenic contamination in groundwater above the permissible limit of 10 µg/L (Ghosh and Singh, 2009). The scientific investigations on the quality of the groundwater and Arsenic occurrence in many other states are tabulated below in Table 2 and Table 3.

Table 2. Districts having arsenic (between 0.01 to 0.05 mg/liter) in groundwater in different states of India

Sl. No.	State	Parts of Districts having As between 0.01 to 0.05 mg/liter
1.	Andhra Pradesh	Guntur, Kurnool, Nellore
2.	Assam	Golaghat, Jorhat, Lakhimpur, Nagaon, Nalbari, Sibsagar, Sonitpur
3.	Bihar	Begusarai, Bhagalpur, Bhojpur, Buxar, Darbhanga, E. Champaran, Gopalganj, Katihar, Khagaria, Lakhisarai, Lohardaga, Madhepura, Muzaffarpur, Purnea, Saharsa, Samastipur, Siwan, Vaishali, W. Champaran
4.	Chhattisgarh	Rajnandgaon
5.	Delhi	East, North East
6.	Daman & Diu	Diu
7.	Gujarat	Amreli, Anand, Bharuch, Bhavnagar, Dahod, Gandhinagar, Kacchh, Mehsana, Patan, Rajkot, Surendranagar, Vadodara
8.	Haryana	Bhiwani, Mahendergarh, Palwal, Rohtak, Sirsa, Sonipat
9.	Himachal Pradesh	Kangra
10.	Jammu & Kashmir	Jammu, Kathua, Rajouri
11.	Jharkhand	Sahebganj
12.	Karnataka	Raichur, Yadgir
13.	Madhya Pradesh	Betul, Burhanpur, Chhindwara, Dhar, Khandwa, Mandsaur, Neemuch, Umaria,
14.	Odisha	Gajapati
15.	Punjab	Faridkot, Gurdaspur, Hoshiarpur, Sangrur, Tarn Taran
17.	Rajasthan	Ganga Nagar
18.	Telangana	Nalgonda
19.	Uttar Pradesh	Azamgarh, Badaun, Bahraich, Basti, Deoria, Gorakhpur, Jhansi, Kausambi, Kushinagar, Maunath Bhanjanm, Pilibhit, Shahjahanpur
20.	West Bengal	Hooghly, Howrah, Kochbihar, Malda, Murshidabad, Nadia, North 24 Parganas, South 24 Parganas

Source: CGWB, 2009.

Table 3. Districts with Arsenic (>0.05mg/liter) in Groundwater in the Different States of India

Sl. No.	State	Parts of Districts having As >0.05mg/litre
1.	Assam	Cachar, Jorhat, Nagaon
2.	Bihar	Godda, Dhanbad
3.	Chhattisgarh	Rajnandgaon
4.	Haryana	Ambala, Jhajjar
5.	Jharkhand	Sahebganj
6.	Karnataka	Raichur
7.	Manipur*	Bishnupur, Thoubal
8.	Punjab	Amritsar, Ropar, Taran taran
9.	Uttar Pradesh	Bahraich, Deoria, Lakhimpur, Azamgarh, Maunath Bhanjan
10.	West Bengal	Hooghly, Malda, Murshidabad, Nadia, North 24 Parganas, South 24 Parganas

*Based on the report of School of Environmental Studies (SOES), Jadavpur University, Kolkata, West Bengal, India

Source: Mukherjee A et al., 2006.

Figure 2. Arsenic in Indian Sub-continent.

The Arsenic has also been reported though in low concentration in Punjab and adjoining states in India. The occurrence of Arsenic is mainly reported from the areas covered by the middle and the lower Ganga basin. Based on the analysis of groundwater samples collected from the tube wells/ Borewells, arsenic hotspots (>0.01 mg/l of BIS) are demarcated in the map of India (Figure 3).

Arsenic is a naturally occurring toxic element that is familiar with it's severe health implications. An excess amount of Arsenic can cause acute gastrointestinal and cardiac damage. Chronic doses can cause vascular disorders such as Blackfoot disease and epidemiological studies have linked Arsenic to the skin and lung cancer.

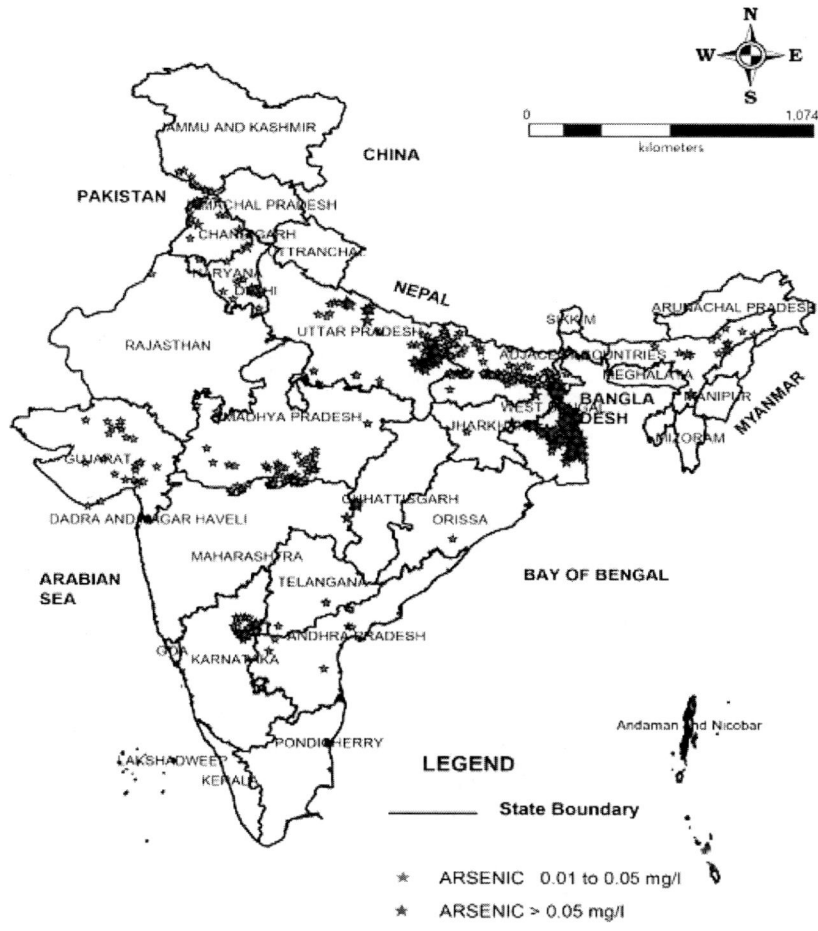

Source: CGWB.

Figure 3. Arsenic hotspots of India.

The arsenic problem in groundwater has emerged as a major quality problem and health hazard in parts of the country within the Indus-Ganga-Meghna-Brahmaputra (I-G-M-B) alluvial plain of the Ganga basin. West Bengal is one of the most arsenic affected states in India. The first case of arsenicosis was reported in the year 1978 in West Bengal. Then, two Kolkata based research institutions, namely School of Tropical Medicine (STM) was confirmed a few cases of arsenic dermatitis and subsequently confirmed by the All India Institute of Hygiene and Public Health (AIIPH).

They jointly established that prolong intake of Arsenic contaminated groundwater was the cause of arsenicosis.

Source: CGWB.

Figure 4. Arsenic contaminated areas of West Bengal.

In 1975, under the Safe Drinking Water Act (SDWA), EPA established the Maximum Contaminant Level (MCL) for Arsenic at 0.05 ppm (U.S. Environmental Protection Agency, 2000). It was observed that the people suffering from Arsenical dermatosis (one of the first symptom of Arsenic poisoning) were consuming groundwater mainly from the hand pump and when the samples of groundwater of the respective area were chemically analyzed; those were found to contain Arsenic beyond the permissible limit of 0.05 ppm. Subsequently, the contamination of groundwater by Arsenic in Bangladesh was also confirmed by the Department of Public Health Engineering (PHED) in Chapai Nawabganj in late 1993, following reports of extensive contamination in the adjoining area of West Bengal. The severe arsenic-contaminated areas of West Bengal is shown in Figure 4.

Drinking water is one of the most important pathways of exposure to Arsenic in the human population and groundwater as a drinking source is thought to be responsible for the majority of the world's chronic Arsenic related health problems. Arsenic contamination in groundwater has emerged as a major quality problem and health hazard in states of the country.

Forms of Arsenic and Their Mobilization

Arsenic has four main chemical forms having oxidation states, -3, 0, +3, and +5, but in natural water, its predominant forms are inorganic oxyanions of trivalent arsenite (As^{3+}) or pentavalent arsenate (As^{5+}) (Smedley and Kinniburg, 2002). The toxicity of different arsenic species varies in the order arsenite > arsenate > monomethyl arsonate > dimethyl arsenate (Vu et al., 2003). Trivalent arsenic is about 60 times more toxic than arsenic in the oxidized pentavalent state, and inorganic arsenic compounds are about 100 times more toxic than organic arsenic compounds (Jain and Ali, 2000). The organic forms of arsenic are quantitatively insignificant and are found mostly in surface waters or areas severely affected by industrial pollution (Smedley and Kinniburg, 2002). The relative concentrations of As(III) to As(V) vary widely, depending on the redox conditions in the geological environment (Jain and Ali, 2000).

The two most important factors controlling the speciation and solubility of arsenic are pH and redox potential. Under oxidizing conditions at pH less than 6.9, $H_2AsO_4^-$ is the dominant species, whereas $HAsO_4^{-2}$ predominates at higher pH. Under reducing conditions at a pH value of less than 9.2, the uncharged arsenite species H_3AsO_3 is dominant (Vu et al., 2003). In contrast to the pH dependency of As (V), As (III) is virtually independent of pH in the absence of other specifically adsorbed anions (Smedley and Kinniburg, 2002). Most often, more trivalent arsenic than pentavalent arsenic is found in reducing groundwater conditions; the converse is true in oxidizing groundwater conditions (Vu et al., 2003). Unlike other toxic trace metals whose solubilities tend to decrease as pH increases, most oxyanions, including arsenate (As^{5+}), tend to become more soluble as pH increases (Vu et al., 2003). When most other metals become insoluble within the neutral pH range, arsenic is soluble at even near-neutral pH in relatively high concentrations. That is why groundwaters are easily contaminated with arsenic and other oxyanions (Smedley and Kinniburg, 2002).

Long-term exposure to low-to-medium levels of arsenic via contaminated food and drinking water can have a serious impact on human health. Globally, more than 100 million people are at risk of arsenic menace. Initiated to address this major health issue, the International Congress on Arsenic in the Environment (ICAE) has been held five times: Mexico 2006, Spain 2008, Taiwan 2010 Australia, 2012, and Argentina 2014. Since the end of the 20th century, arsenic in drinking water (mainly groundwater) has emerged as a global health concern. In the past decade, the presence of arsenic in plant foods, especially ricehas gained increasing attention. Furthermore, in the Nordic countries, in particular, the use of water-soluble inorganic arsenic chemicals (e.g., chromate copper arsenate, CCA) as wood preservatives and the mining of sulfidic ores have been flagged as health concerns. The issue has been accentuated by discoveries of naturally occurring arsenic in groundwater, primarily private wells, in parts of the Fennoscandia Shield and sedimentary formations, with potentially detrimental effects on public health. Sweden has been at the forefront of research on the health effects of arsenic, technologies for arsenic removal, and sustainable mitigation measures for developing countries.

The main objective of this article is to discuss the fundamentals of the arsenic occurrences in terms of its chemical forms and mobility and present a brief review of the existing processes for remediation of arsenic contamination in groundwater. Both ex-situ (on the surface) and in- situ (within the aquifer) remediation processes have been discussed. Ex-situ processes include precipitation, adsorptive, ion exchange, and membrane processes. In-situ processes include immobilization of arsenic by sorption, chemical oxidation and reduction processes, arsenic biotransformation, and arsenic hyperaccumulation in plants. Depending on the magnitude of problems and budget provisions, a suitable method (ex-situ or in-situ) could be chosen for remediation purposes. Natural Attenuation (NA) of arsenic through in-situ processes in soils and groundwater is a very convenient technique, but its performance depends on the prevailing physicochemical and biological processes, which require conducive environmental conditions. Sufficient characterization of the site geology, hydrology, and microbiology is required to model the fate and transport of arsenic. The on-going National Aquifer Mapping (NAQUIM) program by Central Ground Water Board (CGWB) would possibly give a clear insight into the aquifer geometry and arsenic occurrences at the micro-level to derive a suitable remediation technology. Arsenic contamination problems in the context of Punjab State have been highlighted and possible solutions based on studies by Food and Agriculture Organization (FAO), (2007) have been discussed in brief. Research needs are many, and further future studies are necessary.

Geochemistry of Arsenic

Arsenic is a metalloid of group VA element of the Mendeleev's periodic table. Arsenic can exist in oxidation states in the natural environment including +5, +3, +1, and -3 valence and rarely in elemental (neutral) form. In these different oxidation states, Arsenic can form many inorganic and organic compounds. The common valances of Arsenic in unpolluted groundwater of geogenic origin are +3 and +5 as hydrolysis species H_3AsO_3, $H_2AsO_3^{-1}$ $HAsO_3^{-2}$, and AsO_3^{-3} and H_3AsO_4, $H_2AsO_4^{-1}$, $HAsO_4^{-2}$, and AsO_4^{-3}. The dissociation constant of As (III) and As (V) acids is quite different. The organic Arsenic compounds such as Methyl Arsenic acid {$CH_3AsO(OH)_2$} and Dimethyl Arsenic acid {$(CH_3)_2AsO(OH)$} are not common in groundwater. In general, the major processes responsible for the observed concentration of Arsenic in groundwater include mineral precipitation, dissolution, and absorption/desorption, chemical transformation, ion exchange, and biological activity. Under the aerobic condition, pentavalent species predominates while the trivalent state predominates at low Eh in normal water with pH close to neutrality. The redox reaction of Arsenic (III) and Arsenic (V) system;

$$H_3AsO_4 + 2H^+ + 2e = H_3AsO_3 + H_2O \quad E_O = 0.56V$$

If the Nernst equation is applied for the redox potential at pH 7 a value of 0.147V gives an equal concentration of both forms of Arsenic but for 99% oxidation of Arsenic +0.206 V is necessarily indicating early oxidation of As (III) by dissolved oxygen, but the kinetics of the reaction is very slow, at a higher temperature rate increase. Factors such as pH, redox potential, solution composition, competing and complexing ions, aquifer mineralogy, reaction kinetics, and hydraulics of the groundwater system influence the Arsenic concentration in the groundwater. The environment conditions of the sediments have a greater influence on Arsenic speciation and mobility than the total concentration of Arsenic in the sediments. The sand grains in the Arsenic infested aquifers are generally coated with iron and Arsenic rich minerals. In general, the sediments are rich with sand, silt, and clay. The hydrous oxides in sediments occur as a partial coating on the silicate minerals rather than a discrete well-crystallized mineral thus allowing oxide to provoke physical and chemical activities out of proportion to their activity. Under reducing conditions, the sorbed Arsenic from hydrous iron and Manganese oxide is readily mobilized. The role of Sulphides and hydrous Fe, Mn, and Al oxides as important sinks and mode of transport of Arsenic in the environment. Thus the changes in redox potential, acidity, complexing agents can bring about the formation and dissolution colloidal hydrous oxides, and hence that directly controlling the mobilization of Arsenic in an aquatic environment.

SOURCES OF ARSENIC

Arsenic is a natural component of the earth's crust and is distributed at a large scale in the environment in air, water, and land. The inorganic form of arsenic is more toxic than the organic form. Arsenic in the natural environment occurs in soils at an average concentration of about 5 to 6 mg/kg. The high concentration of Arsenic in rocks results from the case with which Arsenic substitutes for Si, Al, or Fe in crystal lattices of silicate minerals. Sedimentary rocks contain higher concentrations of Arsenic compared to igneous or metamorphic rocks (Chadha and Sinha Ray, 1999). Country-wise major contributing sources of arsenic in groundwater are shown in Table 5. Inorganic and Organic arsenic compounds & their chemical bonding in Figure 5. However, anthropogenic sources for Arsenic can be as following:

1. Mining Activities and Smelters - Arsenic is a natural contaminant in lead, zinc, gold, and copper ores and can be released during the smelting process.
2. Pesticide Use - Paris green (Aceto Arsenite) and Pb, Ca, Mg, and Zinc arsenate, zing Arsenite used extensively as pesticides may give rise to Arsenic contamination.
3. Coal - Arsenic can be present in coal mainly as Arsenopyrite at concentrations from <1 to >90 mg/kg.
4. Coal Combustion By-Products (CCBP) - Arsenic in fly ash is derived from coal, which is associated with pyretic material.
5. Sewage Sludge - Arsenic in sewage sludge can be as high as 3 to 46 mg/kg.
6. Irrigation – Crops irrigated by arsenic-contaminated groundwater and food prepared with contaminated water are the responsible sources of exposure.

7. Industrial Processes – The use of arsenic in industries as an alloying agent along with the use of As in processing of textiles, pigments, glass, metal adhesives, ammunition, and wood preservatives are important anthropogenic contributors.
8. Smoking and use of tobacco – Tobacco plants absorb inorganic arsenic from soil and the person smoking tobacco is being exposed to As.

ARSENIC IN THE ENVIRONMENT

Arsenic is a naturally occurring toxic element in rocks, soil, and water. Arsenic in the natural environment occurs in soils at an average concentration of about 5 to 6 mg/kg. The levels of Arsenic in the soils of various countries have been said to range from 1 to 50 ppm (mg/Kg) with a mean of 5 ppm. Arsenic occurs mainly as inorganic species and may be converted to Organoarsenic compound by soil microorganisms. Minerals of Arsenic include Sulphides, Oxides, Arsenites, and Arsenates. Arsenic occurs in a pentavalent state as Arsenic acid and a trivalent state as Arsenite in soil solution and both oxidation states can be subjected to chemically and microbiologically mediate oxidation, reduction, and methylation reactions. Arsenic (III), the reduced state of inorganic Arsenic is a toxic pollutant and it is more soluble and mobile than the oxidized state of inorganic Arsenic, Arsenate [As (V)]. Arsenate can be absorbed onto clays, kaolinite, and montmorillonite at low pH (5.0). Adsorption of As(V) by calcite increased from pH 6 to 10. Organic Arsenic, Dimethyl Arsenic acid (Cacodylic acid) is a volatile Arsenic compound and may be present in all soils. Solubility and speciation of Arsenic in soils is controlled mainly by redox potential and pH. Under oxidizing conditions, Arsenic solubility (200-500 mV) is low, and most (65-98%) As(V) becomes reduced to As(III) and toxic Arsenic is solubilized. Under moderately reducing conditions (0-100 mV), arsenic solubility is controlled by iron oxyhydroxides presence, where Arsenate As(V) is co-precipitated with Iron compounds and released upon solubilization.

Of the total Arsenic added to soils, about 23% comes from clay fly ash and bottom ash, 14% from atmospheric fallout, 10% from mine tailings, 7% from smelters, 3% from agriculture, and 2% from manufacturing, urban and forestry wastes. Arsenic is often associated with Iron Sulphides, especially pyrite since Arsenic can substitute for Sulphur in the pyrite crystalline structure. Coal and CCBP (Coal Combustion By-Product) rich in Sulphur may contain much more Arsenic than those low in Sulphur. The high concentration of arsenic in rocks results from the substitution of Si, Al, or Fe in crystal lattices of silicate minerals. Sedimentary rocks contain higher concentrations of arsenic compared to igneous or metamorphic rocks. Apart from natural sources of arsenic in the environment, anthropogenic sources for arsenic can be mining activities and smelters, pesticide use, coal sewage sludge, etc.

Arsenic retention and release by sediments depend on the chemical properties of the sediments, especially on the amounts of Iron and Aluminium Oxides and Hydroxides they contain. The mechanism of Arsenic retention can be adsorption or co-precipitation. Arsenic release to the environment can be enhanced by subjecting river sediments to oxidation. Arsenic oxidation from As (III) to As (V) may be due to an abiotic process with minor participation of micro-organisms, which is a natural process that helps to alleviate toxicity of As (III) in an aquatic environment, because As (V) is adsorbed onto the sediments and becomes relatively immobilized. Since Arsenic in soils is highly mobile, groundwater is susceptible to Arsenic contamination.

In short, trivalent and pentavalent Arsenic form rather insoluble complexes in sediment systems by interaction with hydrous oxides containing clay particles. Adsorption, desorption, redox potential, and biological transformation reactions influence Arsenic mobility during water-sediment interactions. The adsorption of Arsenic is related to pH, pE, texture, and clay and sand content of sediments.

Arsenic is deposited on sediments mainly as Manganese and iron oxyhydroxides while the Arsenate – Arsenite profile with depth in pore waters is governed by the redox gradient and by the presence of Sulphide. The redox transition zone is defined by bacterial degradation of organic matter and depends on the organic carbon content, the rate of sedimentation and the diffusion of oxygen from overlying water, methylation, demethylation and reduction are important in controlling the mobilization and subsequent distribution of Arsenicals. Organic matter occurring naturally in the aquatic system can interact with Arsenic. Such Arsenic-humic acid interactions at certain pH values may be as important as adsorption to hydroxides.

ARSENIC IN CONSOLIDATED AND UNCONSOLIDATED FORMATIONS

Arsenic in Consolidated Formations

Case Study of Chhattisgarh State, India

This study has highlighted the occurrence and extent of arsenic in Ambagarh Chowki including the health aspects of arsenic in groundwater on local people. A total of 52 and 60 water samples have been collected during pre-monsoon and post-monsoon respectively for analyzing the presence of arsenic in different water sources. The geographic distribution of high arsenic in groundwater is sporadic in occurrence. Of the 52 samples analyzed in May 2017 2 samples have shown arsenic above 50 ppb the BIS permissible limit for arsenic in drinking water. The highest values were 202 ppb in Kaudikasa dug well (near boy's hostel) and 77.94 ppb in the Kaudikasa hand pump. 07 locations have arsenic in the range between 10 ppb and 49 ppb. Those locations are the Telitola hand pump, Joratarai hand pump, Meregaon bore well, Pangri hand pump, Sansaitola hand pump, Kodutola, and Bharritola hand pumps.

The WHO guideline value for arsenic is 10 ppb that means 9 samples above the WHO guideline value. There are 43 samples in which arsenic values were found below 10ppb. Of the 60 samples analyzed in September 2017; 02 samples have shown arsenic above 50 ppb. They are the Taramtola hand pump showing 59 ppb and the Joratarai hand pump showing 53.4 ppb. A total of 05 locations have arsenic in the range of 10 ppb and 49 ppb. Those locations are Kaudikasa dug well and hand pump, Meregaon borewell, Sansaitola hand pump (inside primary school), and Sansaitola hand pump (adjacent to arsenic removal plant). There are 53 samples in which arsenic values were found below 10 ppb. Based on chemical analysis of water samples, it is observed that most of the hand pumps are showing high arsenic content during the pre-monsoon period, which may be due to declining of water level in the hand pump during the pre-monsoon period followed by the release of arsenic from formation to groundwater.

Moreover, it has been observed that the level of arsenic in groundwater is reducing during the post-monsoon period. The reason may be due to the dilution of arsenic in groundwater. But an anomaly from the general trend has been observed in some well locations, where the concentration of arsenic has been exceeded the permissible limit during the post-monsoon period. The anomaly may be due to the dissolution followed by the mixing of arsenic from the topsoil percolating down to the groundwater. Arsenic in dug wells is generally less than the bore well; this may be due to better atmospheric oxidation in dug wells. Groundwater is evaluated to determine its facies by plotting the percentages of chemical constituents in a modified Piper diagram, the groundwater facies is characterized by Ca-HCO$_3$ type covering 69% of the total groundwater samples, and the rest of the samples (31%) are Ca-Mg-Cl type. The ionic dominance pattern among cation is $Ca^{2+} > Na^{2+} > Mg^{2+} > K^+$ and among anions the ionic dominance pattern is $HCO_3^- > Cl^- > SO_4^{2-}$. The arsenic releasing mechanism from host consolidated formation to groundwater through rock-water interaction is shown in Table 4.

Table 4. Arsenic releasing mechanism

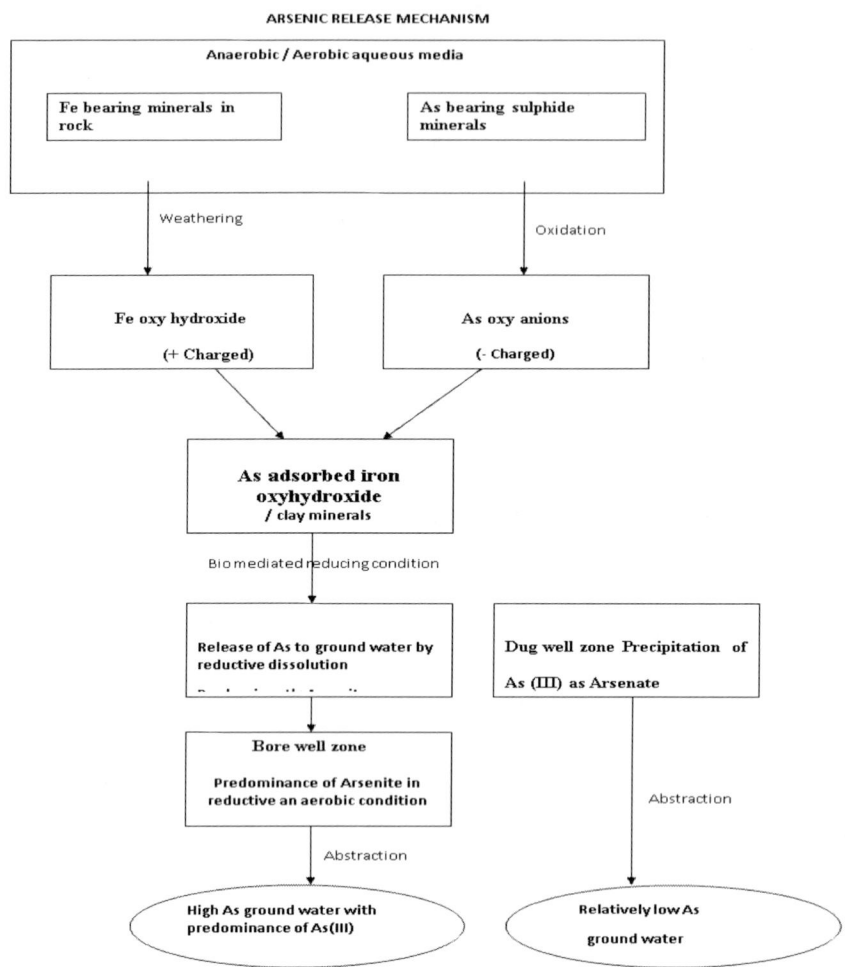

Source: CGWB.

As in Unconsolidated Formations

Groundwater with elevated arsenic affects large tracts of the Bengal basin and is a serious public health hazard (Mandal et al., 1996; Bhattacharya et al., 1997; Nickson et al., 1998). Over 70 million people are exposed to As poisoning in Bangladesh and West Bengal, India, as they depend on As-enriched (>50 ppb) shallow aquifers for drinking water (Burgess et al., 2010). Groundwater in the Ganga Plain, in the upstream of the Garo-Rajmahal Gap, was considered to be relatively As-free, until As enrichment was reported from a few villages on the bank of the River Ganga in the Bhojpur district of Bihar (Chakraborty et al., 2003). Subsequent analyses of spot drinking water sources have revealed As concentrations exceeding 50 ppb in large tracts of the Middle Ganga Plain in Bihar and Eastern Uttar Pradesh (Acharya, 2005; CGWB and PHED, 2005).

Arsenic Issues in Agriculture and Irrigation

It has been estimated that the earth's biomass contains about 30 million tons of arsenic. Hence, plant and animal-based foods are well-recognized routes of arsenic intake. Arsenic in food did not draw much attention, because most of the arsenic (about 75%) present in organic matter as reported in the literature is in less toxic organic form. Some recent studies indicate that the proportion of inorganic arsenic in foods may be significantly high. There are some reports about the high concentration of arsenic in rice produced in Bangladesh. The preliminary findings of UNDP/ FAO studies in Bangladesh showed that the arsenic content of rice produced by irrigation with arsenic-contaminated water is higher, but the relative increase in arsenic content is not significant. The highest concentration of arsenic was found in the roots of the rice plant, then a lower concentration in stem and leaves, and the lowest in rice grain. Arsenic content in all samples tested was within the acceptable limit (0.01 ppm). It appears that arsenic is not readily translocated to the rice grain. Since very high concentrations of arsenic affect the growth of many plants.

It is believed that the rice plant may cease to grow before the rice grain is contaminated with high levels of arsenic. Some plants and bottom-feeding crustacea and fishes used as food, on the other hand, are known to accumulate very high levels of arsenic from the environment in which they grow. The study shows that the concentration of arsenic in arum, a vegetable/food in Bangladesh is 20 mg/kg, which is 400 times higher than the acceptable level in the water. Irrigation with arsenic-contaminated water may contaminate the soil. Since a significant part of irrigation water is lost by evaporation and transpiration, arsenic can accumulate in the root zone like many salts. Conversion of As(III) to As(V) by photo-oxidation in the irrigation field can facilitate such accumulation. Higher arsenic contents in topsoil have been found in many agricultural lands in Bangladesh. The highest concentration so far reported is 83 mg/kg.

The estimated number of excess skin cancer due to arsenic contamination of drinking water in Bangladesh is 415,100. The percentage of incidence of arsenicosis in few Upazilas screened so far is one-third of – that estimated by using the EPA model. Limited data suggest that the skin cancer risk of ten or more for drinking water arsenic content of 0.17 ppb may be an overestimate. On the other hand, the health effects in Bangladesh may be in the preliminary stage, the peak corresponding to the present contamination level is yet to occur. The estimated residual excess skin cancer risks from drinking water supply in Bangladesh complying with Bangladesh Standard and WHO Guideline value are 4.3 and 1.2 per 10,000 populations respectively. Some of the food items can be sources of significant arsenic intake in Bangladesh. Arsenic contaminated groundwater applied for irrigation in parts of India, Bangladesh, and Nepal is responsible for human health threat even causing cancer through eating of that food from crops irrigated by the people of the Indian sub-continent.

The accumulation of poisonous arsenic in the soil layer is posing threat to the future sustainable agriculture in the areas. The arsenic infested areas are to be studied in priority to delineate the contaminated pockets in a micro-scale. After assessing the nature and extension of arsenic toxicity, future suitable site-specific remedial interventions can be designed. Bioremediation methods (Figure 5) through the mixing of fresh vegetables, cow dung, and compost with arsenic-contaminated sludge in the ratio of 1:4 would help to reduce arsenic from the sludge in alarming extent. Cow dung is the most effective arsenic remediation agent followed by fresh vegetables and compost. The optimum period to be maintained for the residential time of the said mixture for removal of arsenic needs to be decided as follow up measures in the future.

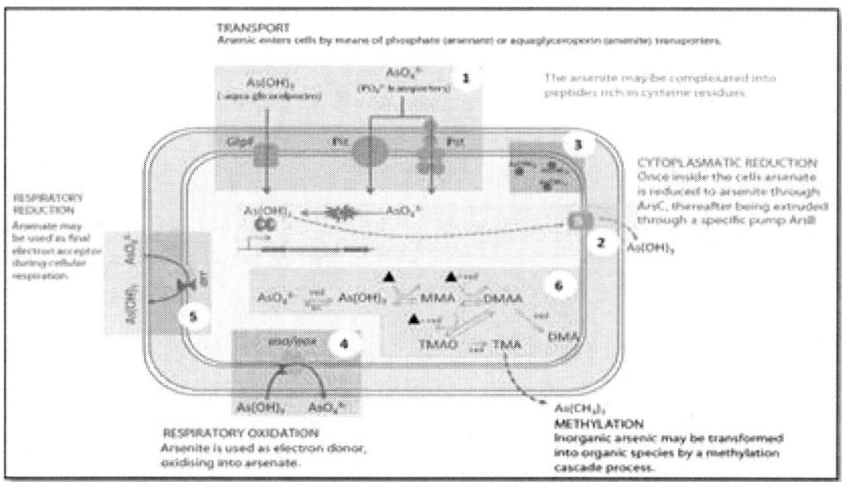

Source: Paez et al., 2009.

Figure 5. Mechanism Involved in Arsenic Bioremediation.

EXPERIMENTAL ARTIFICIAL RECHARGE STUDIES

Hydrogeological Studies conducted by CGWB indicate that mainly the shallow (within 80 metre below ground level) aquifers are contaminated with arsenic. Mainly the tube wells are constructed within arsenic contaminated 80 m bgl depth, which is catering the drinking water need of people residing in the area.

Dilution Study

To monitor intra-aquifer dilution of arsenic, an experiment was carried out at Ashoknagar, in Habra II block of North 24 Parganas district, West Bengal. A shallow tube well (16 metre depth), tapping 12-16 metre of fine sand underlain' different layers of sand and silty clay forming part of an arseniferous aquifer was constructed. A pit of dimension 3.1 metre x 2.8 metre x 2 metre was then constructed around the shallow tube-well. The bottom of the pit (1/2 metre) was filled up gravel to enhance the downward percolation of water. The walls of the pit on the four sides were covered with plastic sheets to prevent the lateral flow of water within the pit. Gravel was also shrouded around the shallow tube well for a recharging effect. An estimated quantity of 47,300 liters of arsenic-free water was charged into the pit during the entire period of the experiment. It was ascertained that the initial concentration of 0.128 ppm of arsenic of the shallow tube-well water was reduced to 0.08 ppm in one month, which finally reduced below the detection limit (BDL) of 0.001 ppm in three months. The entire experiment was conducted during the months from September, 1998 to December, 1998 to ensure that the artificial recharge and not the natural recharge contribute to the dilution exercise.

Impact of Pumping

To monitor the impact of pumping on variation of Arsenic concentration in groundwater, a detailed study was conducted in the Baksha village, Haringhata Block, Nadia district, West Bengal. The study indicates that variation of arsenic concentrations in pumped water with constant discharge oscillates up to three consecutive hours of pumping and there onwards arsenic concentration become stable. The initial fluctuation of arsenic level may be interpreted as pre-energised phase of aquifer systems. The stability of arsenic concentration in groundwater may relate to phenomenon of full-fledged aquifer response against the pumping.

A few hydrological tests were conducted by CGWB to examine the effect of arsenic concentration consequent upon pumping in tube well water. The table (Table5) reflects the pattern in which arsenic concentration in a representative tube well responds to pumping (withdrawal) of water at a constant discharge.

Table 5. Variation of Arsenic concentration with pumping

Location	Depth of the well in metre below ground level (m. b.g.l)	Discharge (m³/hr)	Time for Collection of water samples since pumping started (minutes)	As concentration (ppm)
Baksa, Haringhata Block, Nadia district, West Bengal	19.40	35.00	0	0.55
			30	0.38
			60	0.55
			90	0.56
			120	0.28
			150	0.28
			180	0.21
			210	0.21
			240	0.21

The data presented above reveals that consequent to pumping for 240 minutes, arsenic concentration in tube well water got progressively reduced.

ISOTOPE STUDIES

Case Study in Bengal Delta

A collaborative program between CGWB and BARC (Bhabha Atomic Research Center, Bombay) has been initiated in 1997, and the studies in arsenic - affected areas by isotope techniques are continuing. To study the dynamics of groundwater in arsenic infested areas of Murshidabad, Nadia, South and North 24-Parganas districts in West Bengal, more than one hundred groundwater samples from different depth levels as well as surface water and rainwater were collected during December, 1996, May, 1997, and January, 1998 for environmental isotope ^2H, ^{18}O, ^3H, ^{14}C and a few samples for ^{34}S measurement. Selected samples were measured for δ^2H, $\delta^{18}O$, $\delta^{213}S$, $\delta^{34}S$, ^3H, and ^{14}C, and the results obtained have been summarized below.

Surface Water

A few surface water samples (the Padma, Bhagirathi, and Bhairavi River), were analyzed for δ^2H, $\delta^{18}O$, and ^3H content (Table 6). The results were obtained during December, 1996 and May, 1997 had been compared (Table 6).

The results show that seasonal variation in δ^2H & $\delta^{18}O$. Post-monsoon samples showing more depleted values compared to pre-monsoon water samples. The enrichment in May, 1997 samples are because of greater evaporation of the river water during summer. The tritium values of the river for pre-monsoon and post-monsoon are nearly the same. The arsenic level in the river water is negligible.

Table 6. Isotope results of river water

Sl. No.	Location	$\delta^{18}O$ (%) December, 1996	$\delta^{18}O$ (%) May, 1997	^3H (Tu) December, 1996	^3H (Tu) May, 1997
1.	Padma River	-8.6	-4.47	--	11.6
2.	Bhagirathi River	-8.5	-6.78	10.3	11.9
3.	Bhairavi River	-5.5	-3.53	7.0	6.4

Groundwater Samples

CGWB experiment reveals that in Murshidabad district shallow groundwater δ^2H & $\delta^{18}O$ values are in the range of -46.0% to -24.0% and -8.0 to -3.0 respectively during December, 1996 and -41.0% to -20% and -7.5% to 3.0% respectively during May 1997. These samples fall in the meteoric line on δ^2H- $\delta^{18}O$ plot. The stable isotope results of the shallow groundwater indicate that the recharge to shallow aquifer could be due to precipitation input. Some samples show depleted stable isotopic values during May 1997 (Bichapad $\delta^{18}O$: -7.5%, Ramnagar $\delta^{18}O$: -6.9%) indicating some over river input. Stable isotopic composition of shallow, intermediate, and deep groundwater show similar $\delta^{18}O$ values (6.0%) in Ramnagar and Islampur, showing possible interconnection of two aquifer systems. The tritium values of shallow groundwater generally have 2 to 10 TU showing that they are due to modern recharge. Higher tritium values observed in Lalbagh shallow tubewell, 10.1 TU, Chunakati Nimtala 10.5 TU, Debipur STW 10.6 TU, Bichapad STW 9.0 TU, and Gosakpara could be due to river input. The Carbon-14 data show that the shallow groundwater maintain values in the range of 87 to 109 pmC (percentage of modern Carbon) indicating that they represent mostly modern recharge (<50 years).

The deep groundwater samples from Murshidabad have tritium values >2 TU. ^{14}C values are in the range of 78 to 85 pmC giving age <500 years. Using ^{14}C data, groundwater flow direction, and velocity have been calculated. In Murshidabad district, groundwater flow is from west to east as well as from northwest to southeast direction. There is an indication of groundwater mound formation in the center of the district, and further, it flows towards the south. This is in good agreement with the groundwater contours of that area. Shallow groundwater velocity is in the range of 1.2 to 1.7 cm/d and deep groundwater is about 1.8 cm/d. Variation in groundwater velocity depth viz. could be due to the heterogeneity of sediment. In Nadia district, the stable isotopic composition of shallow, as well as deep groundwater, are more or less similar to that of Murshidabad district. In this district, two shallows and one deep groundwater samples have been measured for Carbon – 14 and the results are 84, 106, and 79 pmC respectively. Water flow is in the North-South direction. The results of isotope studies have been described as below.

In Murshidabad district, shallow groundwater (up to 100m) is due to modern recharge, mostly precipitation recharge. In some parts, the river recharge is also possible.

Deep groundwater contains some components of modern recharge and 14C age may not exceed -500 years. Interconnection between shallow and deep aquifer is possible. In shallow aquifer, groundwater velocity is in the range of 1.2 to 1.7 cm/d and in a deep aquifer, it is about 1.8 cm/d. High arsenic is confined to old river meander belts (eastern part of river Bhagirathi and beyond towards Bangladesh). The Western part of Bhagirathi (older sediment) does not indicate arsenic contamination in groundwater. Persistent clay horizons and fine-grained sandy layers at about 15 to 100 m depth have high arsenic concentration. Arsenic concentration in surface water and deep groundwater is less than the permissible limit (<0.01 ppm). Arsenic content is high in shallow groundwater (<100 metre below ground level depth).

In South 24-Parganas district, Shallow groundwater is a mixture of old and modern water. This aquifer is getting recharge from the river as well as precipitation. Arsenic concentration is high in this shallow aquifer. Deep groundwater is old (^{14}C age: 5000 to 13000 years BP). Interconnection in between shallow and deep aquifer is remote. The reduction of ferric iron seems to be the most plausible mechanism for the release of arsenic into the groundwater of West Bengal. Deep groundwater can be exploited for drinking and domestic purposes only to restrict the further aggravation of arsenic severity.

SPATIO-TEMPORAL DISTRIBUTION AND OCCURRENCES

Investigations by Central Ground Water Board (CGWB) reveals that Arsenic contamination (>0.05 mg/L) is affecting the states of West Bengal, Bihar, Uttar Pradesh, Assam, and Chhattisgarh i.e., Ganga -Meghna-Brahmaputra Plain (GMB plain). The Bengal Delta Plain (BDP) covering Bangladesh and West Bengal in India is the most severe case of groundwater Arsenic contamination. Besides this, high Arsenic groundwater has also been reported from Jharkhand and Manipur state. The occurrence of arsenic in groundwater is sporadic.

SUSTAINABLE ARSENIC MITIGATION IN VARIOUS HYDROGEOLOGICAL TERRAINS

Irrespective of the arsenic concentration of raw water, it is the leachability of sludge containing high arsenic (saturated), which ultimately matters, while deciding on disposal of arsenic-rich sludge safely into the environment without causing any discernible adverse impact on it. Also, total arsenic in the sludge and its leachability needs to be determined, before deciding on its mode of disposal.

Generation of Arsenic Sludge

Arsenic sludge is a toxic waste. Improper disposal of this sludge will contaminate air, soil, water and thus will eventually contaminate the entire ecosystem and environment. In the USA, improper disposal of toxic wastes such as Arsenic sludge is illegal and unlawful, and severe penalties are in imposed for illegal dumping of toxic wastes.

The effective optimization of sludge treatment and disposal system requires correct scientific planning of the operations linking to the treatment steps to those of disposal/ use i.e., storage and transportation. Transportation is a consequence of the fact that the disposal sites are almost invariably far from those of production. It was recognized that coordination collection and safe disposal of Arsenic laden sludge from an individual household, poses a level of complexity and enforcement effort that is difficult to sustain in remote villages.

Techniques for Arsenic Sludge Disposal

Solid wastes management in developing countries is often unsustainable; relying on uncontrolled disposal in the waste dump. Particular problems arise from the disposal residues generated after the treatment of groundwater at the site in a village for drinking purposes, because Arsenic can be highly mobile and have the potential to leach back to the groundwater and surface waters.

Of all the naturally occurring groundwater contaminants, Arsenic is by far the most toxic, and its removal, therefore, must address the consequent disposal and/or containment issues. A national collection and consolidation program is desirable to reduce the potential risks to the environment due to *ad hoc* and uncontrolled long term storage and illegal disposal of Arsenical wastes from the Arsenic removal plants and also from domestic filters. It was recognized that coordinating the collection and safe disposal of Arsenic-laden sludge from individual households poses a level of complexity in regional level and develops an enforcement effort that is difficult to sustain in remote villages. Installation of community-based Arsenic removal units along with a central regeneration facility offers a sustainable solution for Arsenic removal and safe disposal of Arsenic waste in an area.

Arsenic contaminated sludge can be substantially obtained from the treatment of Arsenic contaminated groundwater. Lack of proper management and reuse of this sludge can create further environmental problems as there is a probability of sludge mixing with soil and water. Since As is an element and cannot be destroyed, the best procedure for managing the waste materials is to detoxify the same by changing their chemical structures to a non-soluble state, which is more suited to secure landfill disposal.

It should also be recognized that the quantity of concentrated Arsenical wastes is relatively small and in general, economic reasons must use for existing treatment facilities, wherever possible. The establishment of advanced treatment systems other than stabilization is unlikely to be economically feasible unless contracts are to be let for the treatment of significant quantities of consolidated Arsenical wastes. The sludge generated with high levels of Arsenic cannot be disposed of directly into the landfill due to the risk of underground water contamination through vertical percolation. The high concentration in the leachates requires treatment before disposal to a landfill, which may not be possible because of the large area, where these Arsenic removal plants or filters are located.

As per the guideline of Government of West Bengal, Arsenic sludge generated from Arsenic removal units commissioned by different agencies at different sites is to be safely disposed of so that the As content in the sludge does not constitute a secondary source of contamination either to water or to the environment as a whole. In cases where safe disposal methods and practices have not started yet, the guidelines make it mandatory for the agencies to transfer all the Arsenic sludge generated to a place in Sutahata Block near Haldia in the Medinipur district of West Bengal has been especially disintegrated as Central sludge Disposal Site for the safe up keeping of the sludge.

The treatment and disposal of sludge is an expensive and environmentally sensitive problem. It is also a growing problem worldwide since sludge production will continue to increase. There is no commercial facility operating for the extraction or recovery of Arsenic from wastes for recycling and reuse. Extraction and recovery are unlikely to be commercially available in the future because of the relatively high costs of extraction and recovery and the absence of a market for the product.

Arsenical wastes are difficult to manage as it is not possible to destroy the material in totality. Ultimately Arsenic must be converted to a form that can be safely placed and contained within a landfill or repository. Typically this will involve some form of stabilization, immobilization, and or encapsulation. As such at present, these techniques are the easiest and economically viable option for handling the Arsenic containing sludge from the Arsenic treatment plants in the remote areas. Arsenic can never be destroyed in the environment – it can only change its form. One of the options is to convert inorganic Arsenic through progressively metabolization to organic form as a result of methylation but then Arsenic remains in the environment for potential harm. The following methods are found noteworthy for the treatment of arsenic enriched sludge: Stabilization with Solidification using Cement- Concrete Mixture – a) Sludge mixed with Cement-Concrete Mixture (CCM) and b) Sludge Encapsulated within Cement- Concrete (Encapsulation).

Geomelt Technique

The Geomelt technology is a family of vitrification technologies developed around the *In Situ Vitrification* (ISV) process. The Geomelt processes are commercially available. It is a thermal treatment process that involves the electric melting of contaminated soils, sludge, or other earthen materials, wastes, and debris to permanently destroy, remove, and/or immobilizing hazardous wastes. These techniques may not be economically feasible or be able to process wastes at a reasonable cost unless a large quantity of sludge is available for the treatment.

Ex-Situ Remediation

Most of the ex-situ methods applied to the groundwater extracted from the aquifers are based on the following processes:

- Precipitation processes: coagulation/filtration, direct filtration, coagulation assisted microfiltration, enhanced coagulation, lime/softening, and enhanced lime softening.
- Adsorptive processes: adsorption onto activated alumina (AA), activated carbon, and iron/manganese oxide-based or coated filter media.
- Ion-exchange processes: specifically anion exchange.
- Membrane processes: nano-filtration, reverse osmosis (RO), and electro-dialysis.

In-Situ Remediation

In-situ remediation of arsenic contamination involves natural attenuation (NA) of arsenic concentrations to desired levels. The naturally occurring physicochemical and biological processes suggest the use of NA processes to remediate As-contaminated soils and groundwater. The NA of arsenic mainly involves the following processes:

- Immobilization by sorption.
- Chemical oxidation and reduction.

- Biotransformation.
- Hyperaccumulation in plants.

SOCIO-ECONOMIC SURVEY ON IMPLICATIONS OF HIGH ARSENIC ON HUMAN AND CROP HEALTH

A Case Study in Bengal Delta

The available epidemiological data are evidenced by chronic Arsenical dermatosis. The diseases and disorders so far identified due to drinking excess arsenic over the years are a pigmentation of the skin, keratosis, anemia, respiratory diseases, and hepatomegaly. From the preliminary epidemiological study, it becomes apparent that people drinking Arsenic contaminated water showed clinical symptoms like the pigmentation of the skin, liver diseases, and a few cases of skin cancer. Two major metabolic pathways for Arsenic to enter into the human health system have been identified: i) Oxidation-reduction reactions for the inter-conversion of Arsenites and Arsenates in the body. Methylation reactions that ultimately convert these compounds to Monomethylarsine and Dimethylarsine as metabolic products. Human exposure to either Arsenates (As_3O_4) or Arsenites (As_2O_3) usually results in an increased level of inorganic As (III), As (V), Mono Methyl Arsenate (MMA), and Di Methyl Arsenate (DMA) in the urine. The toxic level of Arsenic causes significant DNA hypermethylation of tumor suppressor genes p16 and p53, thus increasing the risk of carcinogenesis. These epigenetic events have been studied in vitro using human kidney cells in vivo using rat liver cells and peripheral blood leukocytes in humans. The excess Arsenic may cause sufficient damage to human health such as respiratory, cardiovascular, gastrointestinal, hematological, hepatitic, renal, neurological, dermal, and also carcinogenic effects.

Health Surveillance

Thedefensive mechanism in tackling arsenic toxicity depends on early diagnosis and case wise management, detection of arsenic in water, and supply of arsenic-free water. Though state agencies have already initiated many remedial measures, strategies are being undertaken to take care of the future generations, who are threatened by the Arsenecosis. According detailed researches, the best available solution to the problem lies in providing alternative sources of arsenic-free water. A major health hazard has been skin lesions and keratosis sole. However, prolonged exposure leads to infection of the lungs, urinary bladder, and can even cause skin cancer. Further, due to lack of ignorance and lack of information, a two-fold problem arises- i) Complex social issues like stigmatization and non-reporting of early symptoms due to fear of humiliation and isolation; ii) Lack of proper investigation by doctors' resulting in the wrong diagnosis.

A Joint Plan of Action Committee (JPOAC) can be initiated, which will perform with a mandate of three- modes of interventions:

1. Epidemiological survey: It attempts to train doctors, paramedics, and Non-Government Organization (NGO) workers and includes patient identification and counseling of patients about preventive measures. It will help to record the information for health surveillance and assess morbidity patterns.
2. Patient management: Trained doctors on patient treatment strengthen capacities of hospitals and create a database of reported cases. Doctors needs to be trained on the efficacy of drugs and other means of treatment.
3. Study on adverse health implications on women and children: To investigate the effects of arsenic on fertility, pregnancy outcomes, and child development.

RECOMMENDATIONS FOR FUTURE RESEARCH AND SUSTAINABLE MITIGATION STRATEGIES

General Perspectives of Arsenic Remediation

Recommendations

- Priority must be given to solving inequities in access to water as it is closely linked to health and food security.
- Fix standards (Indian Standard-05000) as a mandate in a phased manner and achieve prescribed standards.
- Strengthen overall monitoring.
- Conjunctive use of groundwater, surface water, and roof top rain water harvesting.
- Dual Water supply depending upon the gravity of contaminant.
- Raise public awareness through interpersonal communication related to water, sanitation, health, nutrition, security, hygiene, education Gross Domestic Product (GDP), etc.
- Improvement in case detection and management.
- Improve risk assessment.
- Short term approaches with special reference to the solar system based mini pipe line water supply technology to be given priority.
- Standardization of testing methods on national and global outlooks.
- Concomitant contamination of other heavy metals and bacteria to be addressed on a holistic approach.
- Create sustainable sources in each block.
- Proper training and further studies to address the global problem.
- Regular updating of arsenic monitoring data.
- Each country has comprehensive schemes, but how to strengthen the monitoring strategy, sorting of the risk areas, identifying the population receiving arsenic safe water would be emphasized.

- Standardization of the mitigation technologies available and ensuing the current facilities do not fall back on the performance percentage should be ensured.
- Developing Knowledge, Attitudes, Practices Behavior (KAPB) in the arsenic endemic areas for the creation of awareness generation, knowledge management, capacity building, and formation of a common platform;
- Inter-sectoral coordination and convergence, especially to address the health hazards due to arsenic problems and support health professionals were highlighted.
- Quantifying the arsenic percentage in the food chain and having food safety regulations and standards.
- Following the example of Thailand for 360-degree monitoring by assessing the quality of rainwater.
- Adopting door-to-door campaigning/testing and Information-Education –Communication (IEC) activities as an example set by Cambodia.

Source, Genesis, Migration/Ground Water Security

Recommendations

- Control the water level.
- Recharge the groundwater.
- Catch rainwater.
- Conservation of water.
- Aquifer mapping and identification of vulnerable and safe zone.
- The addition of water especially groundwater goes to sea.
- A revival of local water bodies.
- Monitoring in terms of surface water and groundwater quality.
- Arsenicosis patients should be detected at a very early stage.
- Social awareness for long term sustainability.

- Involve industries for management and research
- A multi-disciplinary and multi-stakeholders nationally coordinated "comprehensive framework of activities" can help to resolve the groundwater Arsenic menace in India.

Arsenic in Food Chain and Food Safety

Recommendations

- Food safety issues to be given maximum attention. Risk assessment is based on food consumption & new safety standards to be formulated.
- Food Safety and Standards Authority of India (FSSAI) limits of metal contaminants in food products to be followed strictly.
- A revision of the standard is expected soon. Specification of Arsenic in various foods is an urgent requirement.
- Use of AMF (Arbuscular Mycorrhizal Fungi), which has a symbiotic relationship with rice that dilutes Arsenic content.
- Growing rice aerobically and use of Molecular Biology Techniques.
- Market Basket (MB) Based Diet Survey techniques need international standardization and India requires studies of Arsenic intake through food based on Total Diet Studies (TDS).
- People living in non-endemic areas are also at risk due to food import from producers from endemic areas. Infants are more at risk due to higher consumption of Arsenic as per body mass.
- Overuse of water due to easier and cheaper access to Deep Tube Well (DTW) should be regulated by the government.
- Entrepreneurship on the ecosystem should come up with a 'Global Sustainability Concept'
- The presence of Arsenic in Egg and Potato is very high— A noticeable issue that has a direct impact on the Food chain.

- Arsenic levels in food grown using contaminated water or cooked using contaminated water both need to be taken into consideration while assessing total Arsenic intake through the food chain.

Technology/Mitigation Options-I

Recommendations

- For getting arsenic-free water, multidimensional/conjunctive use of water for drinking purposes need to be stressed upon-surface water could be the better option.
- Since operation and maintenance from a sustainability point of view are very essential, community participation for long term impact needs to be taken into the consideration – capacity building and institutional development should be the key components of community participation.
- Various models for arsenic mitigation were suggested – the Sujapur Model, Electro-Chemical Arsenic Remediation (ECAR), Arsenic Removal Unit (ARU), arsenic, and horizontal learning program (HLP of Bangladesh), etc.
- Community participation, Local entrepreneurship, and social marketing were emphasized for proper operation and maintenance.

Technology Options/Mitigation Options-II

Recommendations

- The base of the problem lies in the extensive utilization of groundwater for agricultural purposes mainly through Borewell abstraction structures resulting in water level depletion and an

increase in arsenic level. So emphasis needs to be given for evolving new seeds that consume less water to grow.
- The natural (i.e., dry season) and Man-Made difficulties (e.g., on repaired machinery, unsolved social conflicts, etc.) can be overcome by appointing Field Moderators. IEC development, training and survey may be undertaken by the competent authority.
- For the detection of Arsenic, Colorimetric Test kit, Screen-printed electrodes, and for removal Arsenic Ligands silver sensor and Community-scale Resin Based Units are highly efficient.
- Other technology for mitigation of regenerable types may be used.

Socio-Economic and Health Issues and Communication Strategies

Recommendations

- Courtyard education about water safety plant among women, who are mainly chief caretakers and the technologies, should be more women-friendly. As we understand, women are the best water manager.
- Nutritional improvement by supplementing organic selenium can remove the Arsenic from the body.
- Long term assessment for Arsenic in the health system should be emphasized by testing Urine Arsenic and Nail Arsenic samples, clinical changes should be monitored and actively treated.
- Technologies should be used to induce behavioral change amongst people, the audience should be the priority while at the time of imparting the information through.
- Ceramic Membrane Technology (CMT) is recommended as backwash that can be treated efficiently in the system, because disposal of sludge need to be handled much more conveniently.

Human Health Risk and Impact of Arsenic Exposure from Groundwater

Recommendations

- Institution of a standard design and maintenance of a central registry with regular updating and validation of data with regards to arsenic incidence and prevalence.
- WHO neuro-behavioral core test battery design may be used as a screening tool for early detection of arsenicosis: Urinary 8-OHdG may be used as an arsenic-induced cancer marker, along with other ailments like gastrointestinal and respiratory disorders.
- Usage of the Geographic Information System (GIS) model for identifying new arsenic-contaminated areas and linked to health-related changes.
- Socio-economic status. Nutritional and health status. Environmental and exogenous factors, genetic polymorphisms may influence arsenic toxicity.
- Role of Glutathione (GSH). Methyl-donors, Folic acid, Vit-B12, and traditional medicines (Regular in China) have shown co-relation with arsenicosis.
- Nutritional inputs, supplementation needed have to properly assess e.g., using selenium, protein /methylation, etc.
- Comprehensive economic loss needs to be assessed and worked out based on arsenic-related disease load and related issues.
- The entire food safety chain needs to be regularly monitored for absolute measurement of arsenic-related disease burden.

Industry Related Presentations on Technology Options

Recommendations

- Technological opportunities in tracking the migration of heavy metals into groundwater aquifer through simulations exercises are needed.
- Ensure community ownership of water sources through the empowering process.
- Aware the people regarding importance of drinking water quality and relation with health.
- Microbes based sustainable treatment of arsenic and iron contaminated groundwater, which does not generate sludge need to be.
- Granular ferric hydroxide (GFH) based treatment of arsenic for effective removal would be recommended.
- The use of Field Test Kit (FTK) for the detection of arsenic, iron, pH, and chlorine content at the field level was presented.

Monitoring, Surveillance, and Risk Management

Recommendations

- Standardization of sampling techniques, permissible limits, technologies for mitigation, and comprehensive Generalized Weighted Residual Method (GWRM) system was recommended.
- Multi-sectorial approach and collaboration to be considered: Multi-sectorial arsenic mitigation funds to be mobilized.
- Integration of Water Safety Plan with ongoing water quality monitoring and surveillance program and linking with Integrated Disease Surveillance Project (IDSP).

- Disease surveillance system needs to strengthen even in areas where 10 to 50 ppb of arsenic is reported as reports show death rate as high as of 45% of total deaths due to arsenic poisoning without any signs among the populations.
- Due to the other competing priorities in the health system, arsenic is getting low priority.
- Inadequate research in arsenic detection techniques is affecting food safety and security status.

Policies and Implementation Strategies for Arsenic

Policies

- Use rain, surface/ sub-surface safe-water sources as far as possible, conserving consumptive use, refurbishing old surface water sources (ponds lakes, etc.) paying attention to economics.
- Invest in a portfolio of technology options,
  - Innovation for lower cost and more effective technologies;
  - Pilot testing those technologies that have passed stage 1.
  - Implement technologies that are already proven, without waiting for 1 and 2.
- Distinguish between short term and long term policy for arsenic mitigation. Go for multi-stage implementation. The short term aims for full coverage at 50 ppb, long term for 10 ppb.

Implementation Strategies

- Community-scale approach. Not household-scale supply.
- Public-private partnership for implementation, with PHED being the supervisory nodal authority, ensuring quality, quantity, and transparency.

- The community-based approach, with the community's active participation; and demand-based approach with users at the center.

Source, Monitoring, Surveillance, and Risk Management

Recommendations

- Water quality monitoring shall include source coding, depth of well, the color of sources, GPS coordinates culminating into a GIS Layer, twice a year.
- Community involvement and capacity building should be mandatory.
- Standardization of field test kits should be made mandatory and accorded a national standard.
- Arsenic surveillance should include testing of pH, iron, Phosphate, Bicarbonate, manganese, and sulfate.
- Sanitary surveillance and monitoring by the community and remedial actions by the service provider.
- Representative random sampling for an epidemiological study.
- Document all illegal sources.
- Protection of sources through artificial recharge and Insitu dilution of contaminants and monitoring of water levels.
- Household quality supply as an effective risk management tool.
- Community alerts to be in Place for contaminated sources.
- Converge water quality monitoring with Solid and. Liquid Waste Management (SLWM).
- Address the high risk contaminated sources on priority.
- Corrective action on contaminated sources should be monitored in the onlineInstitute for the Management of Information Systems (IMIS).
- Uniform protocol for standardization of labs for creations/ regulation of water sources.

Health and Food Chain

Recommendations

- An Institution of a standard design and maintenance of a central registry with regular updating and validation of data with regards to arsenic incidence and prevalence and their impact is needed.
- Food safety needs to be given maximum attention to control the entry of arsenic in the food chain. Risk assessment based on the food chain. Risk assessment is based on food consumption & new safety standards to be formulated. FSSAI limits of metal contaminants in food products to be followed strictly. A revision of the standard is expected soon. Specification of Arsenic in various foods is an urgent requirement.
- The entire food safety chain needs to be routinely monitored by assessing the entry of arsenic through water used for cultivation as well as through water used for cooking.
- Assess Growing rice aerobically and use of Molecular Biology Techniques.
- Inorganic / Organic Arsenic in food needs to be done.
- People living in non-endemic areas are also at risk due to food imports from endemic areas. Infants are more at risk due to higher consumption of arsenic per body mass. Age*Gender-variations also need to be addressed,
- Pharmacists & health care workers should take active participation in public health programs and by proper training. So that they can take part in the function, Preparation & prevention of the issue.
- WHO neurobehavioral core test battery design may be used as a screening tool for early detection of arsenicosis; Urinary 8-OHdG may be used as an arsenic-induced cancer marker, along with other ailments like gastrointestinal and respiratory disorders.
- Usage of the GIS model for discovering new arsenic-contaminated areas and linked to health-related changes.

- Socioeconomic status, nutritional and health status. Environmental & exogenous factors genetic polymorphisms may influence arsenic toxicity.
- Role of GSH. Methyl-donors, Folic acid, Vit-B12, and traditional medicines (Realgar in China) have shown co-relation with arsenicosis.
- Nutritional inputs, supplementation needed have to be properly assessed e.g., using selenium, protein /methylation, etc.
- Comprehensive economic loss needs to be assessed and worked out based on arsenic-related disease load, SE issues, etc.
- Need to screen cancer patients at an early stage in the Arsenic affected area.

Socio-Economic Issues and Communication Strategies

Recommendations

Investment in water quality and water safety is important at national and state levels. This investment will lead to better economic returns for the Country by reducing disease burden premature deaths and disability and improve the human development index.

- The investments have to flow through institutions like Panchayati Raj linked to people's participation and rural development.
- It is important to recognize that even at the decentralized levels there are groups that are socially excluded. Therefore water quality and safety measures have to pay attention to social inclusion.
- People's awareness is critical not just for technology but also about the related problems. People have to be involved in every step and communication should precede the installation of technological solutions. People have to be involved even in technology development.

- Along with creating awareness behavior change is needed. To bring about behavior change social norms have to be addressed and community mobilization is critical.
- Gender is a cross-cutting factor and attention to gender is critical.
- The creation of a cadre of the core network is critical. A recognized agency or organization has to take ownership and responsibility.

Conclusion

The incidence of arsenic contamination of groundwater used for irrigation as well as for human consumption has assumed the dimension of a large-scale chronic epidemiological problem. Millions of people throughout the world as well as on either side of the West Bengal - Bangladesh border and those living in the extended Ganga-Padma-Bhagirathi delta have been exposed to arsenic pollution and are affected due to slow and long-term ingestion of arsenic compounds. Quite a sizable section among them suffers from varying degrees of Arsenical Dermatosis (ASD). Death has already claimed hundreds of lives, young and old, and many are not even aware of the causative factors behind those diseases. Arsenic toxicity in groundwater has already climbed at the fatality level. Needless to say, the arsenic poisoning of drinking water has already assumed the status of an environmental disaster. The reaction towards such a great calamity has provoked is interesting and needs to be studied closely before we proceed to suggest measures to cope with it. UNICEF and ICEF (India Canada Environment Facility) and other national and world health bodies have responded to this emerging health hazard and many field tests. Various academic bodies, Government organizations, and NGOs are extending support for over a decade to conduct village-level surveys, to evaluate the level of arsenic toxicity in groundwater, to identify the source of arsenic in groundwater, and also to study the impact of arsenic contamination on the health of the people living in affected areas. The initiative has also been undertaken towards epidemiological studies and trained manpower generation followed by the mass awareness in the root level through sensitization to the common people.

The efforts mentioned earlier were certainly well-intentioned but not well-targeted. Hence, the focus of all these flurries of activities was either missing or wrongly placed. The time and money already spent should have by this time resulted in the empowerment of affected people. To achieve this, we may outline the program to take up immediate tasks as follows:

- To form a village committee in each village with a large number of representatives of affected families, in particular, women from such families.
- To arrange interaction with NGOs, Science Forum, or Bodies and Experts and the Village Committees regarding the problem of health hazards caused by arsenic toxicity of drinking water.
- Village committee with the help and technical and financial support of Government orgamizations like PHE department, Panchayati Raj Institutions (PRIs), and NGOs. These organizations would be empowered in all matters such as the method of collection of sample and testing of water, identify each source of contaminated water, monitor the level of toxicity on regular and repeated testing, followed by collecting and keeping the detailed data on groundwater available in this villages.
- The village committee in particular and users in general would have to be adequately and properly trained to reduce the toxicity level of drinking water. Care should be taken to use low-cost technology and resources and expertise locally available. If high-cost water plants are available or installed by any means, resources must be found out to keep machines in order and well maintained.
- The district and sub-divisional hospitals and health centers should be equipped well to treat diseases caused by arsenic poisoning.
- Democratic planning with active involvement of community people in a participatory mode would yield an appreciable outcome in the form of arsenic remedy in the groundwater.

ACKNOWLEDGMENTS

The author conveys his heartfelt gratitude to the Chairman, CGWB, and also to the Member and Regional Director, RGI for according necessary permission to publish this book chapter in the Nova Science Publishers, USA. Special thanks to Dr. AGS Reddy, Retired Scientist of CGWB and to my senior colleague Dr. AVSS Anand, Scientist, RGI for their earnest help and support in preparing this book chapter.

REFERENCES

[1] CGWB (Central Ground Water Board). Concept Note on Geogenic Contamination of Ground Water in India. *Central Ground Water Board*. Ministry of Water Resources. Govt. of India.

[2] Chadha DK & Sinha Ray S. P. (1999) *High Incidence of Arsenic in Ground Water in West Bengal*. Central Ground Water Board, Ministry of Water Resources, Government of India.

[3] Mohan D., Pittman C. U. 2007. Arsenic removal from water/wastewater using adsorbents – a review. *Journal of Hazardous Materials* 142: 1-53.

[4] Smedley Pl, Kinniburgh DG. A review of the source, behavior, and distribution of arsenic in natural waters. *Appl Geochem* 2002;17:517-68.

[5] Mandal BK, Suzuki KT. Arsenic around the world: a review. *Talanta* 2002;58:201-35.

[6] Ahmed F. M. 2001. An Overview of arsenic removal technologies in Bangladesh and India. In: M. Feroze Ahmed. et al., 2001 (Eds). *Technologies for Arsenic Removal from Drinking Water. A compilation of papers presented at the International Workshop on Technologies for Arsenic Removal from Drinking Water*. Bangladesh University of Engineering and Technology, Dhaka, Bangladesh, and the United Nations University, Tokyo. May 2001.

[7] U.S. Environmental Protection Agency (2003) *Arsenic Treatment Technology Evaluation Handbook for Small Systems*. (Report EPA-816-R-03-014, US EPA, Washington, DC).

[8] Ghosh M. M., Teoh R. S. 1985. Adsorption of arsenic on hydrous aluminum oxide. In: *Proc. of Seventh Mid-Atlantic Industrial Waste Conference*, Lancaster, PA, pp. 139–155.

[9] USEPA (the United States Environmental Protection Agency). 2000. Introduction to Phytoremediation. National Risk Management Research Laboratories, *Office of Research and Development*. EPA 600-R-99-107. http://www.clu-in.org/download/remed/introphyto.pdf.

[10] Jain A., Loeppert R. H. 2000. Effect of competing anions on the adsorption of arsenate and arsenite by ferrihydrite, *J. Environ. Qual.*, 29:1422–1430.

[11] Vu K. B., Kaminski M. D., Nuñez L. 2003. Review of Arsenic Removal Technologies for Contaminated Groundwaters. *Argonne National Laboratory*, Argonne, IL, USA. ANL-CMT03/2. http://www.doe.gov/bridge.

[12] FAO (Food and Agriculture Organization). 2007. Remediation of arsenic for agriculture sustainability, food security, and health in Bangladesh. *Water Service*, FAO, Rome, Paris.

[13] Wilkie J. A., Hering J. G. 1996. Adsorption of arsenic onto hydrous ferric oxide: effects of adsorbate/adsorbent ratios and co-occurring solutes, *Colloid Surf.*, A 107: 97–110.

[14] PHED & Ministry of Drinking Water and Sanitation. 2012. *International Conference on water quality with special reference to arsenic*, Kolkata.

[15] D. Chakraborti, M. K. Sengupta, M. M. Rahman et al., "Groundwater arsenic contamination and its health effects in the Ganga-Meghna-Brahmaputra plain," *Journal of Environmental Monitoring*, vol. 6, no. 6, pp. 74N–83N, 2004. View at: Publisher Site | Google Scholar.

[16] *Safe Disposal of Arsenic Rich Sludge from Arsenic Treatment Plants in West Bengal, in India.*

[17] Saha D, Dwivedi SN & Sahu S (2009) Arsenic in Ground Water in parts of Middle Ganga Plain in Bihar- An Appraisal. *Bhu-Jal-Quarterly Journal of Central Ground Water Board* 24 (2-3): 82-93.

[18] Acharyya, S. K. (2005) Arsenic trends in groundwater from Quaternary alluvium in Ganga plain and Bengal basin, Indian subcontinent: insights into influences of stratigraphy. *Gondwana Res.* 8, 55-66.

[19] Burgess, W. G., Hoque, M. A., H. A. Michael, H. A., Voss, C. I., Breit, G. N., and Ahmed, K. M (2010) Vulnerability of deep groundwater in the Bengal Aquifer System to contamination by arsenic, *Nature Geoscience*, 3, 83 – 87.

[20] Nickson R, McArthur J, Burgess W, Ahmed KM, Ravenscroft P, Rahman M. Arsenic poisoning of Bangladesh groundwater. *Nature* 1998; 395:338.

[21] Bhattacharya, P., Chatterjee, D., and Jacks, G., (1997). The occurrence of contaminated groundwater in alluvial aquifers from the Delta Plains, eastern India: options for safe drinking water supply. *Int. J. Water Res. Dev.*, 13, 79–92.

[22] Mukherjee A et al., 2006. Arsenic Contamination in Groundwater: A Global Perspective with Emphasis on the Asian Scenario. *J Health Popul. Nutr.*, 142-163.

In: Groundwater Quality
Editor: Rafael M. Vick
ISBN: 978-1-53618-807-3
© 2020 Nova Science Publishers, Inc.

Chapter 2

GROUNDWATER QUALITY IN THE MEGACITY DHAKA, BANGLADESH: ASSESSMENT AND ENVIRONMENTAL IMPACT

Shama E. Haque[*], PhD
*Department of Civil and Environmental Engineering,
North South University, Dhaka, Bangladesh*

ABSTRACT

Dhaka, the capital of Bangladesh, with a population of over 18 million is one of the fastest growing megacities of the world. The city faces many water resource management challenges that are common to other megacities of the 21st century. Due to rapid urbanization along with increased industrialization, the city struggles to provide sufficient water for its residents. Numerous surface water bodies in and around the city have been subjected to increased contamination and pollution due to indiscriminate disposal of untreated waste by municipal, industrial, commercial and agricultural sources. As such, in an effort to meet the hydrologic needs of its inhabitants, groundwater is being extracted from the Dupi Tila aquifer, which underlies the city. However, the current rate

---

[*] Corresponding Author's E-mail: shama.haque@northsouth.edu.

of groundwater withdrawal is beyond sustainable yields as indicated by a rapidly falling water table. In the past five decades, groundwater pumping in Dhaka has caused groundwater levels to drop more than 61 m and the average declination of static water level is approximately 3 m/year. Abstraction of groundwater at a faster rate than it can be recharged is likely to have negative impacts on the natural environment, including *permanently reduced aquifer storage capacity,* land subsidence, reduction of water in surface water bodies along with deterioration of water quality. This chapter aims to discuss a generalized scenario of groundwater in Greater Dhaka Area focusing on deterioration of groundwater quality over the years and its impact on the environment.

Keywords: groundwater, Dipu Tila aquifer, water quality, megacity, environmental impact

1. INTRODUCTION

Groundwater occurs nearly everywhere below the land surface and subsurface water constitutes roughly two thirds of the freshwater resources of the world. Throughout history, and across the globe, groundwater has been exploited for domestic use, livestock watering and irrigation. Groundwater has also played a crucial role in drinking, irrigation and industrial water demands in Bangladesh over the past five decades (The World Bank 2014; Geen et al. 2002). At present, roughly 97% of the total population of Bangladesh obtain their drinking water from an estimated 10 million tube wells (Escamilla et al. 2011), which tap the shallow parts of alluvial aquifers down to depths of 10–60 m bgl (meter below ground level; DPHE-BGS 2001). Relatively easy accessibility to freshwater from groundwater resources has led to overexploitation of groundwater in the country. In particular, about 80% of the total fresh groundwater withdrawal nationwide is used in its three administrative divisions in the north-central and western zones (i.e., Dhaka, Rajshahi and Rangpur; Figure 1; Qureshi et al. 2015).

Prior to the 1970s, surface water was the primary source of drinking and domestic water supplies in Bangladesh. However, historically, surface water resources in the country have been contaminated with pathogenic microorganisms, causing high levels of mortality among infants and children. Consequently, during the 1970s, the World Bank and the United Nations International Children's and Education Fund (UNICEF) along with the Department of Public Health Engineering, Dhaka, campaigned to tap into the country's widespread groundwater resources what was apparently a safe source of drinking water for the population (Gleick 2000). Groundwater has often been *considered* to be the water source least vulnerable to pathogen contamination as soils have the ability to reduce or eliminate pathogenic microorganisms that move into ground with infiltrating water. By the early 1980s, groundwater was systematically introduced for domestic water supplies, largely by using hand-operated shallow tube wells. Shallow tube wells typically consisting of small diameter tubes (5 cm) that were inserted into the ground at depths of <200 m were considered a relatively easy way of withdrawing safe drinking water from aquifers (Smith et al. 2000, 1093). Since the 1970s, tube wells have become ubiquitous in the country and the number of tube wells is thought to have doubled roughly every five years across rural Bangladesh. While the spread of tube wells has given access to drinking water that is much less contaminated with pathogenic microbes than surface water, the elevated levels of dissolved arsenic (As) commonly found in groundwater has given rise to new health problems. The country is dealing with what has been described as the largest mass poisoning of a population in history due to contamination of groundwater with inorganic As (Smith et al. 2000). Groundwater containing elevated levels of *As* extracted from the shallow aquifers are of natural origin, which has possibly been present in the groundwater for thousands of years (Haque 2018, 210). Elevated levels of As has been detected throughout the floodplain and delta of the Ganges, Brahmaputra, and Meghan rivers, however the delta region of southern Bangladesh is the most contaminated (BGS and DPHE 2001). The adverse human health effects of long-term e*xposure* to As is substantial as As-associated health problems can impact multiple biological systems, at times years following exposure reductions (Naujokas et al. 2013). An estimated 35 to 77 million people in the country have been chronically exposed to As through *drinking* water sourced from groundwater beyond the national health standards of 50 µg/L (Flanagan et al., 2012, 839). However, As is not the only water-quality problem in the Bangladesh's groundwater. In many parts of the country, the groundwater are characterized by elevated levels of dissolved iron (Fe), manganese (Mn) and boron (B). In addition, dissolved uranium (U) concentrations in groundwater appear to be a water quality problem in some areas; however, the precise nature of its occurrence and extent of the problem is poorly understood due to insufficient data (DPHE-BGS, 2001). Moreover, climate change along with rising sea levels are expected to contribute to an increase in salinity in the Bangladesh's coastal region's groundwater systems (Haque 2018, 206). Thus, access to safe drinking water is one of the most important health problems in this predominantly rural country.

Providing safe drinking water to all remains high on the agenda of the Government of Bangladesh. In recent years, in an effort to provide a safe water supply to its populations, the Bangladeshi Government in partnership with internationals donors and local non-governmental organizations introduced a number of technological innovations. Water supply options available in the As affected regions can be brought into two main groups: (i) alternative As-safe water source including piped water supplies, deeper tube wells (> 150 m), improved dug wells, designated safe shallow hand pump tube wells, surface water treatment and rain water harvesting; and (ii) effective filters for As removal and treatment of As contaminated water (Kundu et al. 2016, 254). It is apparent that a low cost and low technology solution to the problem will most likely provide safe drinking water to the As affected population of Bangladesh. Additionally, one of the major goals of the national water policy of Bangladesh is to ensure that all people have access to safe drinking water in the urban areas. Clean and safe water is an important issue in Dhaka, and other densely populated cities across the country, where contamination by microbial pathogens can lead to high rates of diarrhoeal disease, impairing children's growth and overall health.

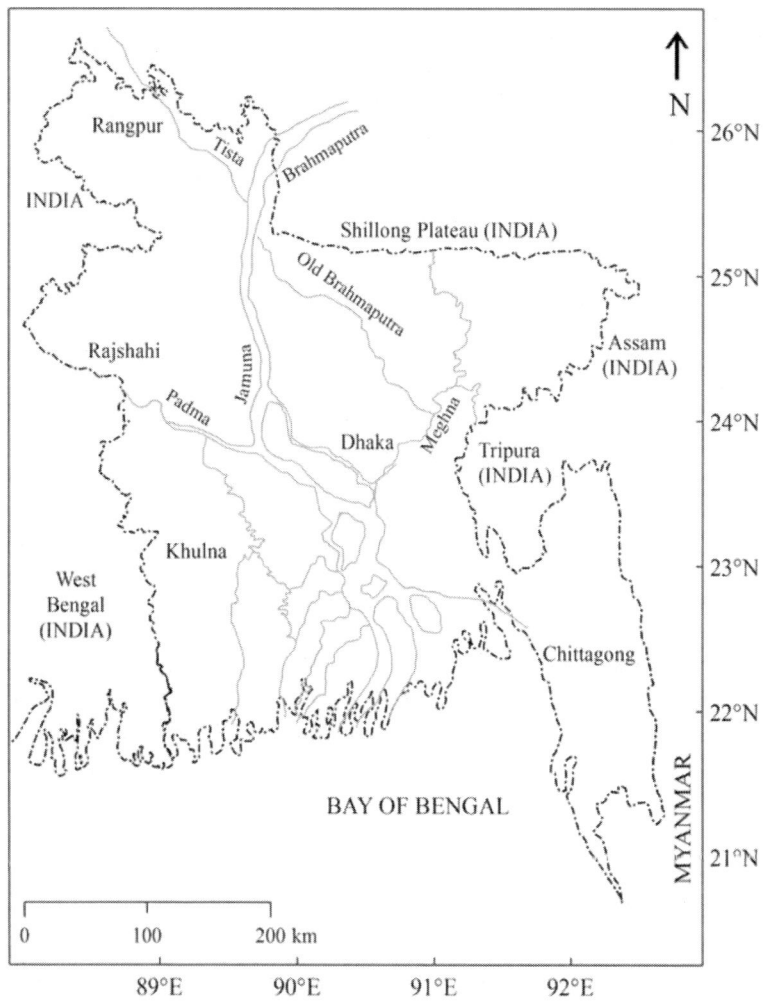

Figure 1. Map of Bangladesh (Modified from Haque 2018).

In the early 1600s, under the Mughal rule, Dhaka was developed on the natural levees of the River Buriganga. Over time, the city has expanded towards a northward direction along the river to accommodate the growing population. Over the past several decades, natural increase of population along with internal mass migration from rural areas across Bangladesh has been the major force behind the booming population of the city. As a result, Dhaka has emerged as one of the fastest growing cities across the globe. Based on the current population growth rate, it is projected that by the year 2025, Dhaka will become one of the top ranking megacities in the world and will be home to roughly 25 million people (Hossain 2011). The rapid growth in population along with the increasing degree of urbanization are the two intertwined reasons that have not only created an enormous stress on the city but also has serious implication for freshwater usage, sanitation and drainage issues, water management issues as well as increasing environmental pollution.

The Buriganga River, which once was considered the lifeline of Dhaka, currently receives a large amount of untreated industrial effluents along with partially treated sewage effluents directly or indirectly from the city. As a result, the Buriganga River is now one of the most polluted rivers in the country (Karim 2005). River-bank infiltration has long been postulated to be a significant source of recharge to the aquifers below the city of Dhaka. Burgess et al. (2011) employed groundwater modelling and environmental isotope distributions independently to determine the importance of possible sources of pollution of the Dipu Tila aquifer that underlies the Dhaka City. Both approaches identified the polluted Buriganga River as the major threat to the quality of groundwater in Dhaka. In addition to the Buriganga River, the city is surrounded by a peripheral river network of roughly 111 km, which includes the Turag, Dhaleshwari, Shitalakshya and Balu Rivers, and the Tongi Khal (i.e., canal). A World Bank report (2009) found that the peripheral rivers of the city are also subject to pollution resulting from surface disposal of untreated industrial and municipal wastewater along with sewage discharge. Furthermore, the city's overall surface water's quality and availability are influenced by seasonal variation depending on the climatic condition of the region. Another major source *of* uncertainty in Bangladesh is that roughly 92% of its surface water originates from out-of-country sources, which gives rise to insecurity in the availability and distribution of surface water (Gain and Giupponi 2014; Rahaman 2009). Due to all these complicating factors, groundwater extracted from the Dupi Tila aquifer, is at present the most important source of water supply in Dhaka (Zahid and Hassan 2017). It is noteworthy that groundwater withdrawn from the aquifer within Dhaka City has very low levels to undetectable levels of dissolved As, a major constraint that limits usage of groundwater for drinking purpose in many areas of the country (Hoque et al. 2007; BGS/DPHE 2001).

The Dhaka Water Supply and Sewerage Authority (DWASA) is responsible for supplying water through pipelines to the city of Dhaka and its neighbouring area. At present, DWASA is capable of producing 25.5 billion litres of water against the daily demand of 24.5-25.0 billion litres (DWASA 2018-2019). Roughly 78% of the water supply system of DWASA comes from groundwater resources using a network of 887 high capacity water wells that tap the lower Dupi Tila aquifer within Dhaka between 150 and 230 m bgl (Hoque et al. 2007; Haque 2006). The remaining 22% of the supplied water comes from the Shitalakshya and Buriganga Rivers, which is treated through four water treatment plants before entering the distribution system.

In 2018, for the first time, Bangladesh fulfilled all three eligibility criteria for graduation from the United Nation's Least Developed Countries list and is on track for graduation in 2024. The rapid economic growth and development of the country has given rise to further demand for energy, transport and urbanization. However, insufficient planning coupled with lack of investment have resulted in increasingly severe infrastructure bottlenecks, overcrowding and environmental degradation. The impact of anthropogenic activities on the evolution of hydrochemistry of the Dipu Tila aquifer is well documented (e.g., Bodrud-Doza et al. 2019; Islam and Huda 2016; Burgess et al. 2011). In and around the city, discharge of untreated industrial effluents and associated toxic compounds into low lying areas has contaminated land and water, and consequently impacted the quality of drinking water and crop products, prompting grave concerns for public health and environment. In a densely populated city as Dhaka, health of millions of people living in the proximity of such areas are affected by the discharged effluent. This chapter aims to summarize existing research conducted on groundwater quality trends of the Dipu Tila aquifer of Dhaka and its effect on human health and the environment.

2. Geological Setting of Dhaka

Geographically, Bangladesh lies between 20°34' and 26°38' N latitude and between 88°01' and 92°41' E longitude. The country shares international land borders with two countries: India to the north, east and west, Myanmar to the southeast and the Bay of Bengal to the south (Haque 2018, 207). The total area of Bangladesh is 147,570 km² and it is located at the lower most reaches of the Ganges-Brahmaputra-Meghna river basin, the largest fluvio-deltaic sedimentary system on Earth. Dhaka City is located in central Bangladesh and lies between latitudes of 23°43' and 23°54' North and longitudes of 90°20' and 90°30' East. The city covers an area of approximately 336 km² (Paul et al. 2019) and it is surrounded by the distributaries of the two major rivers, the Meghna and the Brahmaputra. The peripheral rivers of the city are Tongi Khal to the north, Buriganga River to the south, Turag River to the west and Balu River to the east.

Originally, Dhaka and the adjoining area were built on the southern portion of the elevated tract of the Pleistocene Madhupur Clay Formation, however, with rapid population growth, the city expanded into the flood plain deposits of Holocene age (Alam 1988). In general, the land surface represents uniform geomorphic feature and the topography is flat. The elevation in the central part of the city have an altitude up to 6 to 7 m above mean sea level, whereas the surrounding floodplains are at roughly 4 m above mean sea level. As Dhaka is bounded by the distributaries of several major rivers, the city has been exposed to intermittent flooding since its early days.

3. Climate of Dhaka

Dhaka has a tropical monsoon climate. From a meteorological point of view, the year is divided into four seasons in the city: pre-monsoon (March-May), monsoon (June-September), post-monsoon (October-

November) and winter or dry (December-February; Rahman 2013, 1073). The average temperature in Dhaka is roughly 26°C and the maximum of 33°C commonly occurs in the month of March (Zahid et al. 2006, 496). The temperature during the summer season ranges from 28 and 34◦C and during the winter months, the temperature ranges from 10 to 21◦C (Dewan 2013). The average annual rainfall is about 2076 mm (Ahammed et al. 2014). Majority of the rainfall occurs during the monsoon season, which continues from May till the end of September.

4. Hydrogeological Setting of Dipu Tila Aquifer

The available geological information and relevant research works indicate that the city of Dhaka primarily extracts groundwater from Plio-Pleistocene fluviodeltaic deposits of the Dipu Tila Formation, which comprises of fine to coarse grained micaceous quartzo-feldspathic sand (Burgess et al. 2011). The multi-layered Dupi Tila sand is overlain and confined by the semi-pervious Pleistocene Madhupur Clay, a yellowish brown to oxidised reddish brown silty clay. The *Holocene Alluvial* deposits unconformably overlies the Madhupur Clay, an approximately 48 m thick aquitard (Ahmed et al. 2010, Ahmed et al. 1999). Hasan et al. (1999) report that a comparative analysis of the river bed elevations with the top of the Dipu Tila suggests possible direct contact between the Buriganga River and the aquifer along certain reaches. Additionally, the Madhupur Clay has been reported to be absent in site investigation boreholes drilled at a bridge construction site in southwestern part of Dhaka. Thus, the contact between the peripheral rivers and the Dipu Tila sand is possibly more extensive than previously assumed (Hasan et al. 1999).

The transmissivity of the Dupi Tila aquifer is 500–2,000 m^2/day and the background total dissolved solids content is typically less than 150 mg/L (Hasan and Burgess 1999). The aquifer is relatively homogeneous with respect to aquifer properties and mineral content. However, hydro-structurally the study area is complex as it is bordered by active tectonic units, such as the Sylhet trough in the north, the Jamuna Graben in the west, the Dhaka depression in the south and the Meghna fault zone in the east (Hoque et al. 2007). Additionally, the Madhupur Tract is characterized by numerous faults, which likely influence local aquifer–river systems (Khandoker 1987; Hoque et al. 2007). Historically, regional groundwater flow was governed by boundary faults of the Madhupur Tract and the flow direction was mainly towards the south and southeast of the city (Burgess et al. 2011; DWASA 2006; Ahmed et al. 1999).

5. Trends in Groundwater Levels, Water Quality and Pollution

5.1. Significance of Falling Groundwater Levels in the Dipu Tila Aquifer

During the dry season of the year groundwater levels across Bangladesh become depressed. Typically, groundwater resources are replenished (i.e., recharged) by heavy rainfall and flooding during monsoon season (Haque 2018, 210; Kahlon et al. 2012). However, rates of recharge is variable and controlled by the properties of the overlying soil and aquifer substrate along with methods of recharge. Although the precise recharge mechanism of the Dipu Tila aquifer is yet to be explored, it is well established that *groundwater* development has led to *progressive* water level drawdown in the *aquifer*. Particularly, over the past five decades, groundwater levels have been declining at an average rate of >3 m/year, which has led to substantial drawdown of the Dipu Tila aquifer throughout Dhaka (IWM and DWASA 2011; Akther et al. 2009, 50; DWASA 2006). According to DSCC (2018), the groundwater level is relatively higher in the city's periphery and lower in the centre of Dhaka. Specifically, at the periphery, the northern portion of Dhaka has a relatively higher water level compared to that of the southern portion. The southern portion is more prone to water stress as it is more densely populated and has relatively small surface water bodies that are dispersed throughout the area. Consequently, the rate of groundwater withdrawal is higher in the southern region and due to over pumping, an extensive cone of depression developed in the southern-central area of the city (Ahmed et al. 1998; Hoque 2004).

In recent years, the overall depletion rate of Dipu Tila aquifer has increased as the rates of groundwater abstraction is faster than the rates of natural recharge (Hoque et al. 2007). A key concern of such groundwater depletion is that it can induce increased vertical leakage through areas of contaminated ground both in and around the city (Haque et al. 2013; Bangladesh Water Development Board 2007; Marshall 2005). Progressive groundwater development may also create pumping-induced water pollution in the deeper aquifer by creating a vacuum in the aquifer (Haque et al. 2013). The environmental impacts of groundwater overdraft also include loss of groundwater dependant riparian ecosystems and ground subsidence. Over pumping can lower the water levels below the depth, which riparian vegetation needs to survive (Konikow 2013). Long-term extraction of large amounts of water from a groundwater basin can cause compaction of the subsoil, hence reducing in size and number the open pore spaces in the soil that earlier held water. A study at the Earth Observatory Centre, Dhaka University, found that the city of Dhaka is sinking about 1.27 cm per year on average owing to over extraction of groundwater and lack of recharge of the vacuum it creates in the subsurface. Land subsidence can be associated with other environmental problems including added resistance to natural recharge of aquifers in Greater Dhaka Area and can result in a permanent reduction in the total storage capacity of the aquifer system (DWASA and IWM 2008). In addition, land surface elevation changes can cause serious damages to Dhaka and its neighbouring areas' infrastructures such as roads, bridges, sewers, building foundations, tunnels and pipelines. Preventing land subsidence is therefore of crucial importance. It is also noteworthy that intensive groundwater development has expanded the groundwater management problem from a local issue to one that affects a large region (Khan et al. 2016; Knappett et al. 2016).

5.2. Groundwater Quality

The quality of groundwater in Dhaka is of particular concern as this resource is used for local domestic and municipal water supplies. Groundwater quality is determined by both natural processes (e.g., quality of recharge waters, aquifer lithology and the length of time the water has been in contact with the rocks and subsurface soil) and anthropogenic activities (e.g., discharge of untreated industrial effluent into the environment, agricultural runoff and overdraft of groundwater). To evaluate groundwater quality of Dipu Tila aquifer, Bodrud-Doza et al. (2019, 226) collected and analysed a total of 33 groundwater samples from the aquifer. Hydrogeochemical facies analysis suggest that the studied groundwater are predominantly Ca^{2+}-Mg^{2+}-HCO_3^- type. The findings of this investigation further show that the trends of cations and anions are Na^+ > Mg^{2+} > Ca^{2+} > K^+ and HCO_3^- > Cl^- > SO_4^{2-} > NO_3^-, respectively, and excluding Na^+, K^+, Cl^-, F^- and NO_3^- the remaining reported parameters fall within allowable limits. In addition, silicate weathering was identified to be the dominant process influencing the groundwater solutes. Aqueous speciation simulations *show* that aquifer materials were more saturated with aragonite, calcite and dolomite, and under-saturated with halite.

Bodrud-Doza et al. (2020, 8) found that the Dipu Tila groundwater is slightly acidic to alkaline and dissolved Fe and Mn concentrations exceeded the relevant standard permissible limits. It is well known that human body requires certain essential elements in small quantities, however, high intakes of some of these elements can result in adverse effects. This study further reported that high levels of metal (Mn, Fe, Zn) and anions (F^-, NO_3^-) ingestion from drinking water pose potential health risk to both adults and children living in Dhaka. Additionally, children are especially susceptible to toxic exposures than adults. Children's bodies are in dynamic stages of growth and development, which makes them more vulnerable to harm from toxic exposures. In human body, onset of Mn toxicity depends on the intensity of exposure and on individual susceptibility. Chronic high-level Mn ingestion may lead to neurological disorder and can also be toxic to various organ systems (O'Neal and Zheng

2015). Consumption of excess Fe can cause hemochromatosis, which can cause liver damage, liver cirrhosis, pancreatic islet cell damage, diabetes, hypothyroidism, and hypogonadism (McDowell et al. 2020). Note that compared to several other metal ions with similar chemical properties, Zn is less hazardous to human health. However, elevated levels of dissolved Zn (> 3.0 mg/L) can cause a metallic taste to water (NHMRC 2011).

5.3. Impact of Anthropogenic Activities on the Dipu Tila Aquifer

With rapid urbanization and unplanned growth of the industrial sector many areas across the country are facing widespread contamination of the environment by heavy metals and metalloids. Heavy metals are a naturally occurring group of metals and metalloids that have comparatively high density and are toxic even at very low (i.e., ppb) levels. Examples of heavy metals include Pb, As, Hg, Cd, Zn, Ag, Cu, Fe, Cr, Ni, Pd, and Pt. Heavy metals are considered as one of most significant environmental pollutants due to their persistence in the environment along with their potential for bioaccumulation and its adverse effects for human health. Overall, the degree of environmental contamination by various heavy metals and metalloids vary among cities across Bangladesh, with industrial areas being the most polluted (Islam et al. 2018, 1). Industrial pollution accounts for about 60% of pollution in the Dhaka watershed area, and leather processing along with textile industries are the largest contributors (Ranjan 2019).

5.3.1. Leathering Processing and Its Impact on Groundwater

For example, indiscriminate discharge of untreated effluent from the leather processing industries of Dhaka has resulted in elevated levels of metals in the surrounding rivers (Whitehead et al. 2019). The Hazaribagh industrial area of Dhaka City, located in a highly congested area of less than 30 hectares of land, is the home to roughly 150 leather processing industries (Khan et al. 2020). Roughly 90% of the tanneries in Bangladesh are situated in Hazaribagh, which is situated along the lower reaches of the

Buriganga River. The tanneries have been operating and discharging untreated effluent containing both organic and inorganic compounds, dissolved lime, alkali, chromium sulfate, hydrogen sulfide, sulfuric acid, bleach, dyes, oil, formic acid, various heavy metals (e.g., Fe, S, Mn and Zn), suspended solids, organic matters, etc. directly into the neighbouring areas, river and natural canals (Zahid et al. 2020; Paul et al. 2013, 27; Zahid et al. 2006, 499). It is important to note that globally, The Hazaribagh area of Dhaka is considered one of the hotspots of chromium (Cr) pollution due to the large discharge of untreated Cr enriched effluent over the past several decades. In particular, hexavalent chromium [Cr(VI)] containing compounds are known human carcinogens *(*Norseth 1981*)*. There is general agreement that Cr compounds increase the risk of lung cancer and sinonasal cancer (Nurminen 2005). Many researchers have chemically evaluated the soil and groundwater of Hazaribagh area and reported accumulation of high levels of chromium, aluminum and iron in its top soils (up to 6 m; Zahid et al. 2020; Haque et al. 2019; Zahid and Ahmed 2006). Furthermore, elevated levels of dissolved manganese, zinc, nickel, chromium lead and copper have been detected in shallow groundwater (10–20m) of Hazaribagh area (Zahid et al. 2020; Zahid and Ahmed 2006).

5.3.2. Textile Manufacturing and Its Impact on Groundwater

The textile and ready-made garment industries in Dhaka has a huge water footprint with regards to agricultural water consumption for cotton farming, high water consumption in the manufacturing process and water pollution (Ranjan 2019). Over the past three decades, the export oriented textile and ready-made garment industries of Bangladesh have risen rapidly and have made a considerable contribution to the national economy by creating employment opportunities (Rahman et al. 2007). Based on the growth and its positive impact on the economy, in 2014, the Bangladesh Garment Manufacturers and Exporters Association set an ambitious target for the textile sector to double its exports and reach $50 billion of exports by 2021. According to The International Finance Corporation (IFC) of The World Bank Group, presently, there are 719 washing, dyeing and finishing textile units discharging wastewater into rivers in Dhaka City and the industry is generating nearly 200 metric tonnes of wastewater per tonne of fabric. In general, textile manufacturing processes utilize a huge amount of chemicals for cleaning and dyeing purposes as a result the effluents contain a complex mixture of hazardous chemicals, both organic and inorganic. In Dhaka, majority of the raw or improperly treated effluents are disposed of into neighbouring land, which ultimately ends up in the peripheral rivers that eventually makes its way into shallow groundwater (Sakamoto et al. 2019; Hossain 2018; Dey and Islam 2015, 16).

Textile manufacturing process consumes *a large* quantity of *water* per unit fabric for processing and in Bangladesh, majority of the used water is sourced underground. Textile industries are required to have a permit for groundwater extraction, however, a significant amount of withdrawal is unrecorded due to unregistered or unmetered pumping. The high water demand in textile manufacturing process has resulted in excessive extraction of groundwater, which is reflected by a rapidly falling groundwater table in areas with textile clusters (Hossain et al. 2018). It is noteworthy that according to Sagris and Abbott (2015) export value of Bangladesh's textiles industry will reach to 82.5 billion USD by 2030, which will be accompanied with 250% increase in water demand by the textile sector.

5.3.3. Brick Manufacturing and Its Impact on Groundwater

Brick production contributes 1% to Bangladesh's gross domestic product and generates employment for approximately one million people. Due to ongoing rapid urbanization and associated property development, demand for brick grew steadily at an average rate of 5.6% per year between 1995 and 2005 (UNDP 2010). In Bangladesh, brick remains the most commonly used construction material due to limited supply of natural stone aggregates and other alternative construction materials. Currently, there are approximately 8,000 brick kilns in operation in the country, and in the northern part of Dhaka lies a cluster of roughly 530 brick kilns (Haque 2017). It is well established that brick production results in environmental degradations due to emission of significant quantities of gaseous and particulate pollutants and have been identified to have negative impact on the surrounding areas soil. Soils are the primary sink for heavy metals released into the environment by various anthropogenic activities, including brick kilns. A recent study evaluated the concentrations of various heavy metals (Zn, Cu, Pb, Fe and Mn) in agricultural soil and plant near selected brick kilns in Dhaka (Sikder et al. 2016) and found that soil and plants accumulated maximum amount of micronutrients within 500 to 1000 m distance from brick kilns. The findings of this investigation further suggests the possibility of some of these heavy metals leaching into underground groundwater resources.

5.3.4. Faecal Contamination of Groundwater

The microbiological quality of groundwater is often better than that of surface water as theoretically groundwater is less exposed to and can filter contaminants, particularly faecal bacteria (*Katayama* 2008, 151). Therefore, untreated groundwater is often a traditional means of accessing drinking water in many developing countries. However, microbiological contaminants are detected in groundwater due to poorly constructed wells, septic tanks, landfills and offal holes. Water samples from groundwater supplies in Dhaka show detectable level of Escherichia coli (Ross et al. 2016; Rahman and Quayyum 2008), which is the most commonly used indicator of recent faecal contamination. Escherichia coli pollution has been widely detected in local private tube wells stemming from problems of sanitation. According to The World Bank (2018), in Dhaka, roughly 20% of the population are served by sewage systems, based on official data, however that figure overemphasizes sewage systems' role as the networks frequently malfunction and fail to treat sludge. The remaining faecal sludge from on-site facilities is unmonitored or unregulated. Majority of the population in Bangladesh regularly drinks untreated groundwater and diarrheal disease continues to persist, and accounts for roughly 5% of all deaths in Bangladesh (Saha et al. 2019; Leber et al. 2010).

CONCLUSION

The increased demand for water for Dhaka's growing population has placed an added stress on already stretched groundwater resources. At the local level, drinking water abstraction from Dipu Tila aquifer needs to balance the necessity for using resources versus the effect of over pumping on the environment. If groundwater quality and quantity of Dipu Tila aquifer is to be maintained then effective management of this crucial resource should target to prevent groundwater from becoming severely depleted or highly polluted, since groundwater pollution prevention is always less expensive than remediation.

ACKNOWLEDGMENTS

The author is extremely grateful to Mahbub Haque, Lailun Nahar and Sophie Leila Haque for their unfailing encouragement and infinite patience during the many hours of work dedicated to writing this chapter of the book. The author is especially thankful to her undergraduate students, Samiha Hasan and Alice S. Gomes, for their tremendous *support* during the entire process and particularly for double-checking many details and for preparing the map.

REFERENCES

Ahammed, Faisal, Guna A. Hewa, and John R. Argue. 2014. "Variability of annual daily maximum rainfall of Dhaka, Bangladesh." *Atmospheric Research* 137.

Ahmad, Qazi K. 2001. *Ganges-Brahmaputra-Meghna Region: A Framework for Sustainable Development.* The University Press: Dhaka, Bangladesh.

Ahmed, Kazi M., M. A. Hassan, S. U. Sharif, and K. S. Hossian. 1998. "Effect of urbanization on groundwater regime, Dhaka City, Bangladesh." *Geological Society of India* 52: 229–238.

Ahmed, Kazi M., Makoto Nishigaki, and Ashraf M. Dewan. 2005. "Constraints and issues on sustainable groundwater exploitation in Bangladesh." *Journal of Groundwater Hydrology* 47: 163-179.

Ahmed, Kazi M., Marghoob Hasan, W. M. Burgess, Jane Dottridge, Peter Ravenscroft, and J. J. V. Wonderen. 1999. "The Dupi Tila aquifer of Dhaka, Bangladesh: hydraulic and hydrochemical response to intensive exploitation." In *Groundwater in the Urban Environment: Selected City Profiles*, 19-30. The Netherlands: Balkema, Rotterdam.

Ahmed, Kazi M., Sarmin Sultana, Prosun Bhattacharya, M. A. Haque, and William G. Burgess. 2010. "Groundwater quality of upper and lower Dupi Tila aquifers in the megacity Dhaka, Bangladesh." *Paper presented at Groundwater Quality Management in a Rapidly Changing World (Proc. 7th International Groundwater Quality Conference)*, Zurich, Switzerland.

Akter, Mahmuda, and A. K. M. A. Ullah. 2014 "Water Quality Assessment of an Industrial Zone Polluted Aquatic Body in Dhaka, Bangladesh." *American Journal of Environmental Protection* 3, 232.

Akther, H., M. S. Ahmed, and K. B. S. Rasheed. 2009. "Spatial and Temporal Analysis of Groundwater Level Fluctuation in Dhaka City, Bangladesh." *Asian Journal of Earth Sciences* 2, no. 2 (2009), 49-57. doi:10.3923/ajes.2009.49.57.

Alam, Md. K., and Geological Survey of Bangladesh. *Geology of Madhupur Tract and Its Adjoining Areas in Bangladesh*. 1988.

Bodrud-Doza, Md., Didarul Islam, Tanjena Rume, and Shamshad B. Quraishi. 2020. Groundwater quality and human health risk assessment for safe and sustainable water supply of Dhaka City dwellers in Bangladesh. *Groundwater for Sustainable*. 100374.

Bodrud-Doza, Md., Mohammad A. Bhuiyan, S. M. D. Islam, M. S. Rahman, Md. M. Haque, Konica J. Fatema, Nasir Ahmed, M. A. Rakib, and Md. A. Rahman. 2019. "Hydrogeochemical investigation of groundwater in Dhaka City of Bangladesh using GIS and multivariate statistical techniques." *Groundwater for Sustainable Development* 8: 226-244.

British Geological Survey (BGS) and Department of Public Health Engineering (DPHE). 2001. *Arsenic contamination of groundwater in Bangladesh*. Dhaka: British Geological Survey and Department of Public Health Engineering, Dhaka.

Burgess, William G., Muhammed K. Hasan, Emma Rihani, Kazi M. Ahmed, Mohammad A. Hoque, and William G. Darling. 2011. "Groundwater quality trends in the Dupi Tila aquifer of Dhaka, Bangladesh: sources of contamination evaluated using modelling and environmental isotopes." *International Journal of Urban Sustainable Development* 3: 56-76.

Dewan, Ashraf M. 2013. "Hazards, Risk, and Vulnerability." *Floods in a Megacity*, 35-74.

Dey, Shuchismita, and Ashraful Islam. 2015. "A Review on Textile Wastewater Characterization in Bangladesh." *Resources and Environment* 5: 15-44.

Dhaka South City Corporation, DSCC. 2018. *Dhaka City Neighborhood Upgrading Project (DCNUP) Environmental Management Framework (EMF) Report*. (Government of the People's Republic of Bangladesh Ministry of Local Government, Rural Development and Cooperatives Local Government Division).

Dhaka Water Supply and Sewerage Authority, DWASA. 2019. *Annual Report 2018-2019* (Dhaka).

Dhaka Water Supply and Sewerage Authority and Institute of Water Modelling, DWASA and IWM. 2008. *Final report, resource assessment study, part 2*. (Dhaka Water Supply and Sewerage Authority).

DWASA Resource assessment. 2006. *Final report, vol 1, study conducted by Institute of Water Modelling (IWM)*. Dhaka, Bangladesh.

Escamilla, V., B. Wagner, M. Yunus, P. K. Streatfield, A. V. Geen, and M. Emch. 2011. "Effect of deep tube well use on childhood diarrhoea in Bangladesh." *Bulletin of the World Health Organization* 87: 521 - 527.

Flanagan, Sara, Richard Johnston, and Yan Zheng. 2012. "Arsenic in tube well water in Bangladesh: health and economic impacts and implications for arsenic mitigation." *Bulletin of the World Health Organization* 90: 839-846.

Gain, Animesh, and Carlo Giupponi. "Impact of the Farakka Dam on Thresholds of the Hydrologic Flow Regime in the Lower Ganges River Basin (Bangladesh)." *Water* 6, no. 8 (2014), 2501-2518. doi:10.3390/w6082501.

Geen, Alexander V., Habibul Ahsan, Allan H. Horneman, Ratan K. Dhar, Yan Zheng, Iftikhhar Hussain. 2002. "Promotion of well-switching to mitigate the current arsenic crisis in Bangladesh." *Bulletin of the World Health Organization: the International Journal of Public Health 2002* 80: 732-737.

Gleick, Peter H. 2000. *The World's Water 2000-2001: The Biennial Report on Freshwater Resources*. Washington: Island Press.

Haque, Mahfuzul. 2017. "Promoting environment-friendly brick kilns in Bangladesh: opportunities and challenges." *Dhaka University Journal of Development Studies* 2: 147-159.

Haque, Shama E. 2018. ''An Overview of Groundwater in Bangladesh.'' In *Groundwater of South Asia*, edited by Abhijit Mukherjee, 205-232. Singapore: Springer Publishing Company.

Haque, Shama E., Nazmun Nahar, Alice S. Gomes, Samiha Hasan, Ahsan Saif, Ali S. Sakib, and Sadia T. Nezum. 2019. "The Impacts of Partial Relocation of Hazaribagh Leather Processing Industries on the Environment and Human Health: Focus on Children and Vulnerable Population." *Paper presented at International Symposium on Environment, Disaster and Health, 'Health at Risk in the World of Degraded Environment and Disaster.'* The Institute of Disaster Management and Vulnerability Studies, University of Dhaka.

Haque, Syeda J. 2006. *Hydrogeological characterization of the lower Dupi Tila Aquifer of Dhaka City*. 50.

Haque, Syeda J., Shin-ichi Onodera, and Yuta Shimizu. 2012. "An overview of the effects of urbanization on the quantity and quality of groundwater in South Asian megacities." *Limnology* 14: 135-145.

Hasan, M. K., William Burgess, and Jane Dottridge. 1999. "Impacts of Urban Growth on Surface Water and Groundwater Quality." *Paper presented at Proceedings of IUGG 99 Symposium HS5*, Birmingham.

Hassan, M. Q., and Anwar Zahid. 2011. "Management of overexploited Dhaka City aquifer, Bangladesh." *Journal of Nepal Geological Society* 43: 277-283.

Hoque, Mohammad A. 2004. "Hydrostratigraphy and aquifer piezometry of Dhaka City." *Institute of Water and Flood Management, BUET, Dhaka*.

Hoque, Mohammad A., M. M. Hoque, and Kazi M. Ahmed. 2007. "Declining groundwater level and aquifer dewatering in Dhaka metropolitan area, Bangladesh: causes and quantification." *Hydrogeology Journal* 15: 1523-1534.

Hossain, Shahadat. 2011. *Urban Poverty in Bangladesh: Slum Communities, Migration and Social Integration*. London: I. B. Tauris.

Hossain, Laila, Sumit K. Sarker, and Mohidus S. Khan. 2018. "Evaluation of present and future wastewater impacts of textile dyeing industries in Bangladesh." *Environmental Development* 26: 23-33.

Islam, M. M., Md. Karim, Xin Zheng, and Xiaofang Li. 2018. "Heavy Metal and Metalloid Pollution of Soil, Water and Foods in Bangladesh: A Critical Review." *International Journal of Environmental Research and Public Health* 15: 2825.

Islam, S. M., and M. E. Huda. 2016. "Water Pollution by Industrial Effluent and Phytoplankton Diversity of Shitalakhya River, Bangladesh." *Journal of Scientific Research* 8: 191-198.

Institute of Water Modelling and Dhaka Water Supply and Sewerage Authority IWM and DWASA. 2011. "Establishment of Groundwater Monitoring System in Dhaka City for Aquifer Systems and DWASA Production Wells, Draft Final Report." *Institute of Water Modeling and Dhaka Water Supply and Sewerage Authority*.

Karim, Mir F. 2005. "A note on some geological advantages for construction of underground railway transit system in the city of Dhaka." *Geological Survey of Bangladesh*.

Katayama, Hiroyuki. 2008."Detection of Microbial Contamination in Groundwater." In *CSUR-UT Series: Library for Sustainable Urban Regeneration*, 151–169. Japan: Springer.

Khan, Abidur, Nils Michelsen, Andres Marandi, Rabby Hossain, Mohammed A. Hossain, Karl E. Roehl, Anwar Zahid, Muhammad Q. Hassan, and Christoph Schüth. 2002. "Processes controlling the extent of groundwater pollution with chromium from tanneries in the Hazaribagh area, Dhaka, Bangladesh." *Science of the Total Environment* 710: 136213.

Khan, Mahfuzur R., Mohammad Koneshloo, Peter S. Knappett, Kazi M. Ahmed, Benjamin C. Bostick, Brian J. Mailloux, Rajib H. Mozumder, Anwar Zahid, Charles F. Harvey, Alexander van Geen, and Holly A. Michael. 2016. "Megacity pumping and preferential flow threaten groundwater quality." *Nature Communications* 7: 1.

Khandoker, R. A. 1987. "Origin of elevated Barind-Madhupur areas, Bengal basin results of geotectonic activities." *Bangladesh Journal of Geology* 6:1–7.

Knappett, P. S. K., B. J. Mailloux, I. Choudhury, M. R. Khan, H. A. Michael, S. Barua, D. R. Mondal, M. S., Steckler, S. H. Akhter, K. M. Ahmed, B. Bostick, C. F. Harvey, M. Shamsudduha, P. Shuai, I. Mihajlov, R. Mozumder, and A. van Geen. 2016. "Vulnerability of low-arsenic aquifers to municipal pumping in Bangladesh." *Journal of Hydrology* 539: 674-686.

Konikow, Leonard F. 2013. "Groundwater depletion in the United States (1900−2008)." *Geological Survey Scientific Investigations Report 2013−5079*, 63.

Kundu, Debasish K., Bas J. Van Vliet, and Aarti Gupta. 2016. "The consolidation of deep tube well technology in safe drinking water provision: the case of arsenic mitigation in rural Bangladesh." *Asian Journal of Technology Innovation* 24: 254-273.

Leber, Jessica, M. M. Rahman, Kazi M. Ahmed, Brian Mailloux, and Alexander Van Geen. 2010. "Contrasting Influence of Geology on E. coli and Arsenic in Aquifers of Bangladesh." *Ground Water* 49,111-123.

Marshall, Jessica. 2005. "Megacity, mega mess." *Nature* 437: 312-314.
McDowell, Lisa A., Pujitha Kudaravalli, and Kristin L. Sticco. 2020. *Iron Overload*. StatPearls Publishing, 2020.
Morris, Brian L., Ashraf A. Seddique, and Kazi M. Ahmed. 2003. "Response of the Dupi Tila aquifer to intensive pumping in Dhaka, Bangladesh." *Hydrogeology Journal* 11: 496-503.
National Health and Medical Research Council (NHMRC). 2011. "Australian Drinking Water Guidelines 6." *National Water Commission: Canberra, Australia.* 3: 1-1142.
Naujokas, Marisa F., Beth Anderson, Habibul Ahsan, H. V. Aposhian, Joseph H. Graziano, Claudia Thompson, and William A. Suk. 2013. "The Broad Scope of Health Effects from Chronic Arsenic Exposure: Update on a Worldwide Public Health Problem." *Environmental Health Perspectives* 121: 295-302.
Norseth, Tor. 1981. "The carcinogenicity of chromium." *Environmental Health Perspectives* 40: 121–130.
Nurminen, Markku. 2005. "Overview of the Human Carcinogenicity Risk Assessment of Metallic Chromium and Trivalent Chromium." *The Internet Journal of Epidemiology* 2:1.
O'Neal, Stefanie L., and Wei Zheng. 2015. "Manganese Toxicity upon Overexposure: a Decade in Review." *Current Environmental Health Reports* 2: 315-328.
Paul, Hira, Paula Antunes, Anthony D. Covington, P. Evans, and P. Philips. 2013. "Bangladeshi Leather Industry: An Overview of Recent Sustainable Developments." *Society of Leather Technologists or Chemists*, 25-32.
Paul, Siddhartho S., Syed H. Akhter, Khaled Hasan, and Md. Z. Rahman. 2019. "Geospatial analysis of the depletion of surface water body and floodplains in Dhaka City (1967 to 2008) and its implications for earthquake vulnerability." *SN Applied Sciences* 1: 6.
Qureshi, Asad S., Zia U. Ahmed, and Timothy J. Krupnik. 2015. "Groundwater management in Bangladesh: An analysis of problems and opportunities." *An analysis of problems and opportunities Cereal*

Systems Initiative for South Asia - Mechanization and Irrigation (CSISA-MI).

Rahaman, Muhammad M. 2009. "Integrated Ganges basin management: conflict and hope for regional development." *Water Policy* 11: 168-190.

Rahman, Md. T., Md. Habibullah, and Md. A. Masum. 2007. "Readymade Garment Industry in Bangladesh: Growth, Contribution and Challenges." *IOSR Journal of Economics and Finance* 08: 01-07.

Rahman, Md M., and Shahriar Quayyum. 2008. *Sustainable Water Supply in Dhaka City: Present and Future*. Bangkok, Thailand: Conference of the Japan Science and Technology Agency.

Rahman, Mohammad A., Bettina A. Wiegand, A. B. M. Badruzzaman, and Thomas Ptak. 2013. "Hydrogeological analysis of the upper Dupi Tila Aquifer, towards the implementation of a managed aquifer-recharge project in Dhaka City, Bangladesh." 21: 1071-1089.

Sagris, Thomas, and Justin Abbott. 2015. *An analysis of industrial water use in Bangladesh with a focus on the textile and leather industries*. 2030 Water Resources Group.

Saha, Ratnajit, Nepal C. Dey, Mahfuzar Rahman, Prosun Bhattacharya, and Golam H. Rabbani. 2019. "Geogenic Arsenic and Microbial Contamination in Drinking Water Sources: Exposure Risks to the Coastal Population in Bangladesh." *Frontiers in Environmental Science* 7: 57.

Sakamoto, Maiko, Tofayel Ahmed, Salma Begum, and Hamidul Huq. 2049. "Water Pollution and the Textile Industry in Bangladesh: Flawed Corporate Practices or Restrictive Opportunities?" *Sustainability* 11: 1951.

Sikder, Abdul H., Sayma Khanom, Md F. Hossain, and Zakia Parveen. 2016. "Accumulation of Zn, Cu, Fe, Mn and Pb due to brick manufacturing in agricultutral soils and plants." *Dhaka University Journal of Biological Sciences* 25: 75-81.

Smith, Allan H., Elena O. Lingas, and Mahfuzar Rahman. 2000. "Contamination of Drinking-Water by Arsenic in Bangladesh: A

Public Health Emergency." *Bulletin of the World Health Organisation* 83: 177-186.

United Nations Development Programme, UNDP. 2010. "Improving Kiln Efficiency in the brick making Industry." PIMS#2837.

Whitehead, P. G., G. Bussi, R. Peters, M. A. Hossain, L. Softley, S. Shawal, L. Jin, A. P. N. Rampley, P. Holship, R. Hope, and G. Alabaster. 2019. "Modelling heavy metals in the Buriganga River System, Dhaka, Bangladesh: Impacts of tannery pollution control." *Science of the Total Environment* 697: 134090.

The World Bank (International Bank for Reconstruction and Development). 2018. *Promising Progress a Diagnostic of Water Supply, Sanitation, Hygiene, and Poverty in Bangladesh.*

The World Bank. 2014. *Benchmarking to Improve Urban Water Supply Delivery in Bangladesh. Water and Sanitation Programme Report.*

Zahid, Anwar, and Syed R. Ahmed. 2006. "Groundwater research and management: integrating science into management decisions." 27–46. India: *Proceedings of IWMI-ITP-NH International workshop on creating synergy between groundwater research and management in South and Southeast Asia.*

Zahid, Anwar, K. D. Balke, M. Q. Hassan, and Matthias Flegr. 2006. "Evaluation of aquifer environment under Hazaribagh leather processing zone of Dhaka city." *Environmental Geology* 50: 495-504.

Zahid, Anwar, M. E. Uddin, and F. Deeba. 2004. "Groundwater level declining trend in Dhaka city aquifer." *International workshop on water resources management and development in Dhaka City*, 133.

BIOGRAPHICAL SKETCH

Shama E. Haque

Affiliation: North South University

Education: PhD in Environmental Science and Engineering (The University of Texas at Arlington, USA), BS in Civil Engineering (The University of Texas at Austin, USA)

Research and Professional Experience: Dr. Shama E. Haque is an environmental scientist and engineer with over twelve years of research and professional geochemistry experience. Dr. Haque has worked on natural surface and groundwater systems as well as on systems that have been impacted by mining industry in Canada. She has extensive experience in investigating and evaluating the fate and transport of contaminants in sediments, surface water and groundwater resources. Her expertise also includes studying the geochemical evolution of arsenic along flow paths in several well characterized aquifers in the U.S., and developing conceptual model of arsenic mobility and speciation along flow paths based on field and laboratory studies.

Professional Appointments: Dr. Haque was appointed to the faculty of North South University in 2016, where she currently serves as an Associate Professor in the Department of Civil and Environmental Engineering. In addition, Dr. Haque serves as an Associate Editor for the Journal of Groundwater for Sustainable Development (Elsevier). Dr. Haque completed her postdoctoral study of groundwater evolution in natural and contaminated aquifers at the University of British Columbia, Canada.

Publications from the Last 3 Years:

Haque, Shama Emy. 2018. "An Overview of Groundwater in Bangladesh." In *Groundwater of South Asia*, edited by Abhijit Mukherjee, 205-232. Singapore: Springer Publishing Company.

In: Groundwater Quality
Editor: Rafael M. Vick
ISBN: 978-1-53618-807-3
© 2020 Nova Science Publishers, Inc.

Chapter 3

FROM RAINWATER TO GROUNDWATER CHEMISTRY. CASE STUDY: MT. CAMEROON AREA

Kengue N. J. Dirane[1], MSc, Andrew A. Ako[2], PhD, Fozao K. Folepei[3], PhD, Bertil Nlend[2], PhD and Gloria E. T. Eyong[2], PhD

[1]University of Buea, Cameroon
[2]Hydrological Research Centre Yaoundé, Cameroon
[3]University of Bamenda, Cameroon

ABSTRACT

Rainfall may contribute to surface and groundwater sources. In this study rainwater and groundwater sampling was done and analyzed for chemical composition in the Mt. Cameroon area for the period May-July 2017.

Water samples were investigated for their physico-chemical characteristics and assess the impact of water-rock interactions on groundwater chemistry in the Mt. Cameroon area. The wide ranges of EC values (3.1-367µS/cm) and total dissolved solids (1.74-108.03mg/L) revealed the heterogeneous distribution of hydrochemical processes within the groundwater of the area. The relative abundance of major cations and anions in the analyzed water (mg/L) is $Ca^{2+}>Na^+>Mg^{2+}>K^+>NH_4^+$ and $HCO_3^->Cl^->SO_4^{2-}>NO_3^-$, respectively. All the groundwater samples were soft, with total hardness values (2.54-136.65 mg/L) below the maximum permissible limits of the World Health Organization (WHO) guideline. A Piper diagram classified the water types into Ca-HCO$_3$ type (67%), Na-Cl type (17%) and the rest from the Na-Ca-HCO$_3$ and Ca-Mg-Cl water types; indicating recharge, seawater intrusion, and mixed respectively. Alkaline earth metal contents dominated those of alkali metals in most of the samples. Based on ion geochemistry, water-rock interactions, mixing, ion exchange and anthropogenic activity are the dominant hydrogeochemical processes. The weathering type corresponds to bisiallitization indicating the genesis of smectites where kaolinite and montmorillonite are the main minerals formed. A reasonable conclusion is that the groundwater chemistry is dominated by them and rainwater of local origin.

Keywords: rainwater, groundwater, chemistry, Mt. Cameroon

1. INTRODUCTION

Rainwater is a dilute solution having the lowest ionic content of all natural waters. Its composition results from various processes involving the atmosphere, hydrosphere, and lithosphere. Rainwater acquires its chemical characteristics through dissolution of the gases and solids with which it comes into contact with during its descent. Since human activities can also influence rainwater chemistry, it can be said that rainwater reflects the rain-air-rock-life interaction. Thus, rainwater chemical composition analyses can be used to answer environmental problems resulting from air quality conditions and these conditions vary from one location to another because of local influences.

Therefore, rainwater composition plays an important role in scavenging soluble components from the atmosphere and helps us to understand the relative contribution of different sources of atmospheric constituents (Kulshrestha et al., 2003).

Volcanic areas are very peculiar hydro-environmental systems where rainfall dynamics and volcanic atmospheric emissions are in close and mutual relationships. Rainfalls on volcanoes are chemically influenced by plume-rain interaction processes (Madonia and Liotta, 2010). Volcanic degassing represents an important source of atmospheric gases and aerosol (Oppenheimer, 2003; Mather et al., 2004). During rain events, interaction between water droplets and volcanic products gives rise to rainwater that is chemically marked by the dissolution of soluble ions and gases (Liotta et al., 2006).

Groundwater is an indispensable resource to sustain livelihoods. Its chemical composition is the result of the composition of water that enters groundwater reservoirs and the reactions with minerals present in the rock that may modify the composition. Nearly all groundwater originates as rain that infiltrates through the soil into flow systems in the underlying geologic materials. Water moving through the ground will react to varying degrees with the surrounding minerals (and other components), and it is these rock-water interactions that give the water its characteristic chemistry (Morgenstern et al., 2009, Mohapatra et al., 2012).

This work presents information on the chemical characteristics of rain and groundwater in the Mt. Cameroon area which strongly depends on its atmosphere and the underlying material. Taking into consideration: a) the region's location on the Cameroon Volcanic Line (CVL) as one of the main watersheds in Cameroon; b) the importance of rainfall chemistry in identifying pollutant sources. Rainfall is among the primary source of recharge for groundwater in the Mt. Cameroon area, it is derived from recharge occurring during the rainy season. A better evaluation of groundwater resources in the Mt. Cameroon area is a strategic point for water resources management in Cameroon.

2. MATERIALS AND METHODS

2.1. Description of the Study Area

2.1.1. Location of the Study Area

The study area is located in the south-west part of Cameroon in Central Africa. It is mainly characterized by the Mt. Cameroon area and its environs (Figure 1). Mt. Cameroon lies on the coast, in the Gulf of Guinea, between 3°57'- 4°27'N and 8°58'-9°24'E. The main peak is at 4°7'N, 9°10'E at an altitude of about 4,100 amsl. It lies along the CVL which stretches from the Gulf of Guinea through the western High Plateau, Lake Nyos, and the Oku Massif right up to the Ngaoundéré Plateau.

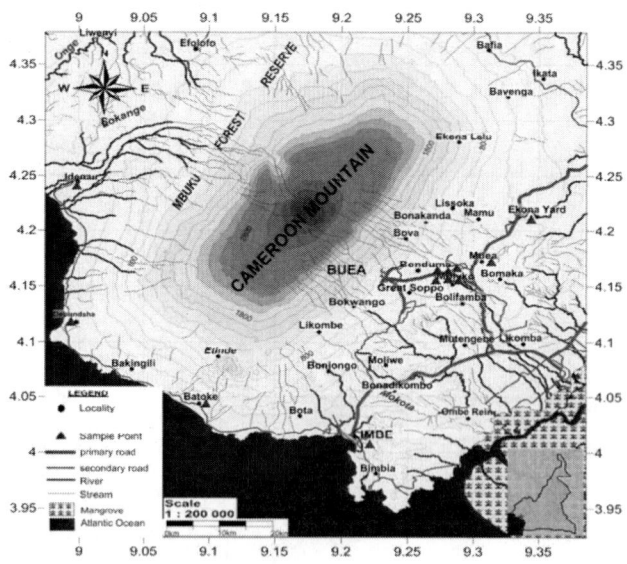

Figure 1. Map of study area showing the sampling points (Base Map from National Institute of Cartography, Cameroon, 2013).

2.1.2. Geology of the Study Area

Mt. Cameroon area is a lava dominated volcano characterized by a rift zone eruption system. This area is made up of rocks both of plutonic and volcanic origin. The plutonic rocks are made up of granites, and syenites to which intermediate and basic rocks are associated. The volcanic rocks include basalts, basanites and/or nephenelites to trachytes, rhyolites and/or phonolites (Njonfang et al., 2011). Figure 2 below shows the geology of the Mt. Cameroon region especially Buea on its eastern slope.

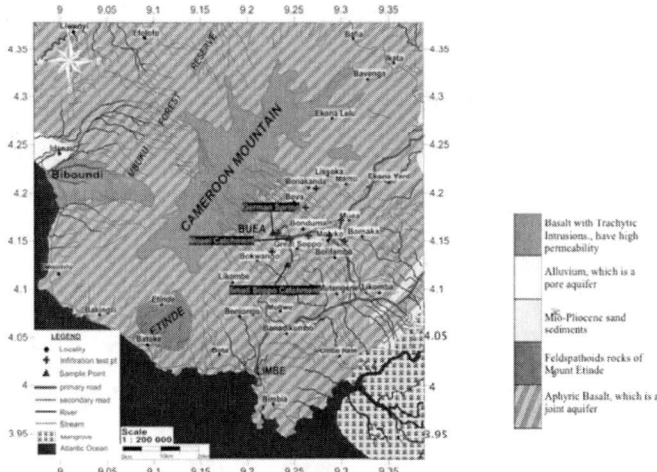

Figure 2. Geological Map of Buea including Mt. Cameroon (Base Map from National Institute of Cartography, Cameroon, 2013).

2.1.3. Climate

Most of this region has a distinct seasonal pattern of rainfall, related to the north-south movement of the Inter-tropical Convergence Zone (ITCZ). The climate of the Mt. Cameroon area is characterized by its seasonal nature; the seasons are very well defined. There is a period of heavy rains occurring between the months of June and October, and a dry period extending from November to May.

The temperature falls with increasing elevation. For each 100m ascent, the average temperature drops by about 0.6°C. The temperature on the top of Mt. Cameroon would be a chilly 4°C when in Limbe at the base it is 32°C. Tchouto et al., (1998) described a "zone of greatest mistiness" between 1,200m and 2,000m, especially from 1,500m to 1,800m, and considered that the highest overall humidity on the mountain occurs there.

2.1.4. Hydrogeology

In spite of the abundant precipitation, the Mt. Cameroon area is characterised by a low hydrological network with a number of rivers, seasonal and permanent streams, springs, lakes and waterfalls. Main rivers include Lokange, Mokoko, Ebié and Onge, most of which empties into the Atlantic Ocean. Most of the rivers, streams are found at the lower slopes of the mountain. There are no permanent rivers/streams at the upper slopes of the mountain. Key waterfalls are located in Bomana, Mbakossi, Ekoumbe, and Ombe localities.

2.2. Sampling and Analytical Methods

Water samples were collected from several sources between May and July 2017 within 8 localities in the study area. The sources comprised 'boreholes' (hereafter termed groundwater), streams, and rainfall events, a total of 12 samples were collected. Field measurements and water sampling followed methods described by Karklins (1996). Collected water samples were properly capped and preserved in a cooler container containing ice blocks prior to analysis, which was undertaken as soon as possible.

The different physical parameters pH, EC, TDS and alkalinity of sampled water were measured in situ using a multi-parameter EXTIC II brand.

The collected samples were analysed in a laboratory institution at the Institut de Recherche et de Geologie Minieres (IRGM), Yaounde after two weeks of sample collection. Major cations: Na^+, K^+, Mg^{2+} and Ca^{2+} were determined by a Flame Technique in a High-Resolution Continuum Source AAS (ContrAA 700) as described by Welz et al., (2006). The anions: F^-, Cl^-, NO_3^-, SO_4^{2-}, PO_4^{3-} were determined by using an Ion Chromatography (Dionex ICS-900). The reliability of the chemical analyses was verified by using an ionic balance error (IBE) equation (Appelo and Postma, 2005). The values were within ±5%; hence, suitable for geochemical interpretations.

3. RESULTS AND DISCUSSION

The water samples collected were divided into 6 Rainfall water sample (RWS), 3 Surface water sample (SWS), and 3 Groundwater sample (GWS).

Statistical information of the measured parameters are summarized in Table 1. The value of these statistical parameters are first presented considering all the sampling points and then stratifying the data by the source categories.

Table 1. Statistical summary of the physico-chemical data of water samples in the study area

PARAMETER	Water Samples (n=12)				Rainwater (n=6)				Groundwater (n=3)				Surface (n=3)			
	Min	Max	Mean	SD	Min	Max	Mean	SD	Min	Max	Mean	SD	Min	Max	Mean	SD
pH	4.24	8	6.64	1.22	4.24	7.59	5.85	1.26	6.69	8	7.37	0.66	7.23	7.7	7.47	0.24
EC (μS/cm)	3.1	367	180.73	144.94	3.1	356	106.6	140.39	212	367	308.67	80.75	59.10	321	201.03	132.32
Alkalinity (μE/l)	0.0	1319	439.39	404.98	0.0	607	182.45	277.24	592	752	665.67	80.75	227	1319	727	551.78
Na$^+$ (mg/l)	0.08	9.36	4.28	3.45	0.08	9.04	2.56	3.42	4.78	9.36	7.11	2.29	1.27	8.44	4.88	3.58
NH$_4^+$ (mg/l)	0.0	0.30	0.08	0.13	0.0	0.0	0.0	0.0	0.0	0.30	0.14	0.15	0.0	0.0	0.0	0.0
K$^+$ (mg/l)	0.04	1.91	0.91	0.75	0.04	1.91	0.49	0.73	1.16	1.83	1.58	0.37	0.51	1.67	1.09	0.58
Mg^{2+} (mg/l)	0.0	4.67	1.52	1.86	0.06	4.67	0.92	1.82	0.0	3.95	2.48	2.16	0.0	3.68	1.78	1.84
Ca^{2+} (mg/l)	0.18	12.87	6.49	5.17	0.18	12.06	3.66	5.19	8.02	12.87	10.39	2.43	2.88	11.51	8.24	4.69
F$^-$ (mg/l)	0.03	0.64	0.18	0.15	0.03	0.64	0.20	0.23	0.12	0.17	0.15	0.02	0.17	0.21	0.19	0.02
Cl$^-$ (mg/l)	0.09	6.79	2.99	2.08	0.09	4.53	2.32	1.84	2.57	6.79	4.77	2.11	0.93	4.97	2.57	2.12
NO$_3^-$ (mg/l)	0.0	26.48	5.04	7.74	0.0	12.51	3.16	4.99	1.74	26.43	12.15	12.79	0.92	2.52	1.70	0.80
PO$_4^{3-}$ (mg/l)	0.0	0.0	0.0	0.0	0.0	0.0	0.0	0.0	0.0	0.0	0.0	0.0	0.0	0.0	0.0	0.0
SO$_4^{2-}$ (mg/l)	0.29	35.14	5.06	9.66	0.29	3.69	1.91	1.43	1.57	35.14	14.51	18.06	0.88	3.93	1.90	1.76
HCO$_3^-$ (mg/l)	0.0	80.46	26.8	24.7	0.0	37.03	11.13	16.91	36.11	45.87	40.61	4.93	13.85	80.46	44.35	33.66
TH (mg/l)	0.98	49.29	22.46	18.79	0.98	49.29	12.90	19.47	20.05	48.37	36.14	14.55	13.96	40.92	27.89	13.50
TDS (mg/l)	1.74	108.03	53.35	41.93	1.74	70.55	26.42	30.89	65.80	108.03	93.89	24.32	23.09	107.45	66.67	42.25
RE	4.96	5.28	5.09	0.14					4.96	5.28	5.14	0.12	4.97	5.18	5.04	0.12

Max: maximum; Min: minimum; SD: Standard Deviation; EC: Electrical Conductivity; TH: Total Hardness; TDS: Total Dissolved Solids, RE: Weathering Index.

3.1. Physical Characteristics

The pH values for the 12 water samples ranged from 4.24 to 8, with an average of 6.64±1.21 which is within the weakly acidic-alkaline range. As to pH data, the distribution of pH shows that 50% of the samples had a pH < 7.

The values for EC ranged from 3.1 - 367 μS/cm with a mean value 180.73 μS/cm. Those for TDS of all water samples ranged from 1.74 -

108.03 mg/l with a mean of 53.35 mg/l. The EC and TDS values of all water samples were low with mean values of 180.73 µS/cm and 53.35 mg/l, respectively suggesting low mineralized and freshwater. The alkalinity (Alk) of the samples varied between 0.0 and 1319 mg/L, and had an average value of 439.39 mg/L ($n = 12$).

The range of water pH values reveals that the samples were generally weakly acidic-alkaline. The slightly acidic nature of the water sources is linked to the formation and dissolution of minerals and also influenced by biochemical processes in solution (Freeze and Cherry, 1979; Williams and Benson, 2010; Nduka and Orisakwe, 2011). Also, this slightly acidic nature of the water sources may also lead to the breakdown of HCO_3^- releasing free hydrogen and carbondioxide. Ako et al., (2011) and Fantong et al., (2009) obtained similar findings in groundwater sources along the CVL and attributed the presence of dissolved CO_2 to root respiration, decay of organic matter or dissolution of carbonate minerals.

The EC of the analyzed water samples in the study area varies between 3.1-367µS/cm with a mean of 180.73±144.94 (Table 1); this represents water experiencing slight mineralization (Gnazou et al., 2011). The increasing mean EC values recorded in groundwater compared with rainwater and surface water samples reflect significant water-rock interaction resulting in the dissolution of geological medium i.e., Mean EC for GWS>SWS>RWS as follows 308.67 µS/cm, 201.03 µS/cm, 106.6 µS/cm respectively.

For our analyzed water samples, the values of TDS were between 1.74 and 108.03mg/L these values were lower than the range accepted by the WHO standards, according to Freeze and Cherry (1979) represent fresh water. It can be observed that the mean TDS increased progressively from a minimum in rainwater (26.42 mg/L) to surface water (66.67 mg/L) and a maximum in groundwater (93.89 mg/L), this indicates the influence of the sediments and bedrock material in chemically enriching the water as it percolates through the aquifer.

Despite their relatively neutral pH values within the WHO guideline value, TDS and EC exhibited a wide range. The high standard deviation of 41.93 and 144.94 for TDS and EC, respectively, indicates the heterogeneity of the hydrochemical processes evolving within the water of the study area (Ako et al., 2011; Gountié et al., 2017).

3.2. Chemical Characteristics

From Table 1 above, it clearly shows that the mean concentration of ionic species in the water samples follows the order of: $HCO_3^- > Ca^{2+} > SO_4^{2-} > Na^+ > NO_3^- > Cl^- > Mg^{2+} > K^+ > NH_4^+ > F^- > PO_4^{3-}$. Among all the ions, the mean highest contributing ions are HCO_3^- and Ca^{2+} for the anions and cations respectively (Figure 3). Together, they account for 62% of the total amount of the ions. The rest of the ions are present in the order stated above and are presented below, showing the dominant presence of HCO_3^- and Ca^{2+} in all the water types.

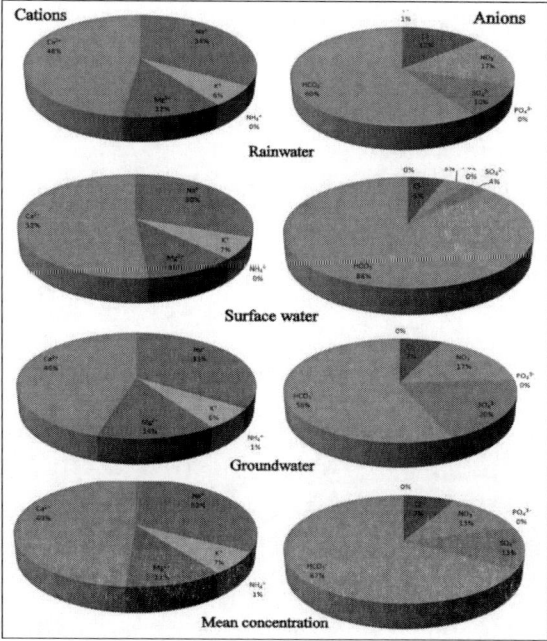

Figure 3. Pie charts of mean concentrations of ions (mg/l) for a) rainwater, b) groundwater, c) surface water and d) summary of mean constituents showing dominant Ca^{2+} and HCO_3^- in the water samples.

3.2.1. Cations

In rainwater, the concentrations of cations were as follows $Ca^{2+}>Na^+>Mg^{2+}>K^+$ with 48% Ca^{2+} present. The cations concentrations in the rainwater decreased with a mean concentration of 12.06, 9.04, 4.67, and 1.91 mg/L, respectively.

For the groundwater, Ca^{2+} (46%) still dominates the cations concentration as follows $Ca^{2+}>Na^+>Mg^{2+}>K^+>NH_4^+$ with mean concentrations of 10.39, 7.11, 2.48, 1.58, 0.14 mg/L respectively.

Concerning the surface water, Ca^{2+} and HCO_3^- had well over 50% of the ions present in the order $Ca^{2+}>Na^+>Mg^{2+}>K^+$ for the cations with mean concentrations of 8.24, 4.88, 1.78, 1.09 mg/L.

3.2.2. Anions

Anions in the rainwater were in the order $HCO_3^->NO_3^->Cl^->SO_4^{2-}>F^-$ with 60% HCO_3^- present. Their concentrations decreased with a mean concentration of 37.03, 12.51, 4.53, 3.69 and 0.64 mg/L, respectively.

Whereas for the anions in the groundwater, it is $HCO_3^->SO_4^{2-}>NO_3^->Cl^->F^-$ with mean concentrations of 40.61, 14.51, 12.51, 4.77, 0.15 mg/L respectively.

Surface water concentrations were $HCO_3^->Cl^->SO_4^{2-}>NO_3^->F^-$ with mean values of 44.35, 2.57, 1.90, 1.70, 0.19 mg/L respectively.

The relative abundance of cations and anions (Figure 3) were as follows:

3.2.2.1. Cations

The values of these parameters in the analyzed water ranged between 0.18 and 12.87; 0.00 to 4.67; 0.08 to 9.36; 0.04 to 1.91 mg/L respectively for Ca, Mg, K and Na. All measured parameters had values well below the recommended WHO standards.

The relative abundance of the cations is as follows $Ca^{2+}>Na^+>Mg^{2+}>K^+$, this shows that Ca^{2+} is the most abundant cation followed by Na^+ in the water samples of the study area. Their main source maybe from the weathering of calco-sodico feldspaths and hornblende in basic rocks (Magha et al., 2015) as explained below.

2Hornblende + 30H$^+$ + 9H$_2$O → Kaolinite + 12H$_4$SiO$_4$ + 4Ca^{2+} + 10Mg^{2+} + 2Na$^+$ (1)

The most common source of K$^+$ is the silicate minerals: orthoclase, nepheline, leucite and biotite. The basement rock of the study area is composed of granite and the K$^+$ is probably from the dissolution of orthoclase (KAlSi$_3$O$_8$) and FeMg minerals like biotite (K (Mg Fe)$_3$ (AlSi$_3$)O$_{10}$ (OH, Fe)$_2$ present in the granites.

2Orthoclase + 2H$^+$ + 9H$_2$O → Kaolinite + 4H$_4$SiO$_4$ + 2K$^+$ (2)

2Biotite + 13H$^+$ + CO$_2$ + 2H$_2$O → Kaolinite + 4H$_4$SiO$_4$ + 6Fe^{2+} + 6Fe^{3+} + HCO$_3^-$ + 2K$^+$ (3)

K$^+$ concentrations in volcanic rocks are usually low, due to the resistance to weathering of potassium-bearing minerals and its fixation in the formation of clay minerals. With the lavas in this area forming a basanite series composed of minerals such as clinopyroxenes, olivines (forsterite) and plagioclase. Hence, the principal cations in the area are calcium and magnesium ions (Djieto et al., 2017).

Albite in the granites is the probable source of the Na$^+$

2Albite + 2H$^+$ + 9H$_2$O → Kaolinite + 4H$_4$SiO$_4$ + 2Na$^+$ (4)

Meanwhile the main source of NH$_4^+$ is probably from fertilizers used in farms which is being leached into the water while natural sources such as organic (metabolic processes) and inorganic (rock weathering and hydrothermal activity) could also contribute to the presence of NH$_4^+$.

Similar observations of silicate dissolution in granitic formations (Njitchoua and Ngounou-Ngatcha, 1997; Fantong et al., 2009) and volcanic terrains (Tanyileke et al., 1996; Ako et al., 2012) along the CVL have been reported with the following generalized equation (Tanyileke et al., 1996).

Rocks + H_2O + CO_2 → Cations + H_4SiO_4 + HCO_3^- + solids (mostly clay minerals) (5)

3.2.2.2. Anions

The relative abundances of anions (mg/l) in the water samples were as follows: $HCO_3^- > SO_4^{2-} > NO_3^- > Cl^- > F^-$. All the samples had values below the WHO standards. HCO_3^- concentrations were high with average values ranging from 11.13-40.61mg/L. The dominance of HCO_3^- is consistent with most natural waters along the CVL. The primary source is the dissolved CO_2 in rainwater and the decay of organic matter at the surface which may release CO_2 for dissolution (Tanyileke et al., 1996, Ako et al., 2012, Wirmvem et al., 2013, Magha et al., 2015, Gountié et al., 2017). Weathering, precipitation and organic decay probably account for the dominance of this ion:

CO_2 + H_2O → H^+ + HCO_3^- or CO_2 + H_2O + CO_2 → $2HCO_3^-$ (6)

SO_4^{2-} could be derived from natural sources such as sulphate minerals common in igneous rocks (Magha et al., 2015). On the other hand, the possible source of SO_4^{2-} in the atmosphere can be derived from SO_2 in the air as well as dry deposition of particles over the study area. It can also be from the oxidation of pyrite:

FeS_2 + $2O_2$ + H_2O → $2SO_4^{2-}$ + H_2. (7)

The concentrations of NO_3^- in the water samples suggest their potential source may be from pollution by agricultural entrants (rainwater), urine, shallow pit toilets and oxidation of organic matters in biochemical processes (groundwater). Agriculture and pollution can be used to explain these levels.

The chlorides in water are known to come from rainwater. Cl^- concentration of 0.09mg/L and 6.79mg/L were obtained in the study area which might have been due to disinfection through chlorination.

The likely source of F⁻ in the granitic basement is fluorite, CaF_2, the most common F⁻-bearing mineral (Edmunds and Smedley, 1996). High F⁻ concentrations in groundwater have been identified in North Cameroon, along the CVL, in a granitic alkaline environment from fluorite, amphiboles and micas (Fantong et al., 2009). The low concentrations of F⁻ in groundwater of the study area may be due to its acidic nature which renders F⁻ immobile (Hem, 1989; Edmunds and Smedley, 1996, Wirmvem et al., 2013).

The mean relative concentrations of dissolved ions show a progressive increase from the lowest in rainwater, through surface water, and the highest in groundwater (Figure 3). These results suggest chemical evolution from rain to surface water and groundwater.

The low major ion concentrations in groundwater depict low water-rock interactions in the granitic basement, short residence time, the shallow nature of the aquifer and its acidic nature (Edmunds and Smedley, 1996). Similar low pH and major ions have been reported from granitic basement complex aquifers in Ghana (Adomako et al., 2011) and Nigeria (Edet et al., 2011).

In volcanic terrains (such as the CVL) the solute composition of waters is principally controlled by the hydrolysis of silicate rock minerals coupled to the slightly acidic pH of water (Ako et al., 2011; Lu et al., 2015; Djieto et al., 2017).

3.3. Origin of Major Ions

3.3.1. Correlation Factors

In order to identify the possible relationship between the various ionic species in water samples, linear regression analysis was carried out. Table 2 gives the Pearson correlation coefficients computed from the 12 collected samples. Strong correlation was seen between some species; they are highlighted in the correlation matrix below.

Table 2. Pearson's correlation coefficients for physicochemical parameters of water samples (Important relationships are highlighted in bold)

Parameter	pH	Cond	Alc	Na⁺	NH₄⁺	K⁺	Mg²⁺	Ca²⁺	F⁻	Cl⁻	NO₃⁻	SO₄²⁻	HCO₃⁻	TH	TDS
pH	1														
Cond	0,657	1													
Alc	0,665	**0,829**	1												
Na⁺	0,695	**0,912**	**0,783**	1											
NH₄⁺	0,218	0,518	0,126	0.362	1										
K⁺	**0,774**	**0,905**	**0,774**	**0,983**	0,424	1									
Mg²⁺	0,468	0,665	0,256	**0,539**	**0,537**	**0,556**	1								
Ca²⁺	**0,768**	**0,968**	**0,857**	**0,927**	0,368	**0,918**	**0,648**	1							
F⁻	-0,153	-0,127	-0,106	-0,089	-0,180	-0,132	-0,084	-0,145	1						
Cl⁻	0,136	**0,530**	0,350	0,421	-0,038	0,364	0,473	**0,539**	0,008	1					
NO₃⁻	0,261	0,372	0,266	0,470	-0,259	0,406	0,191	0,435	-0,116	**0,685**	1				
SO₄²⁻	0,050	0,488	0,285	0,134	**0,536**	0,177	0,428	0,329	-0,128	0,469	0,027	1			
HCO₃⁻	0,665	**0,829**	**1,000**	**0,783**	0,126	**0,774**	0,256	**0,857**	-0,106	0,350	0,266	0,285	1		
TH	**0,718**	**0,936**	**0,693**	**0,856**	0,471	**0,857**	**0,852**	**0,951**	-0,134	**0,563**	0,377	0,401	**0,693**	1	
TDS	**0,646**	**0,939**	**0,918**	**0,840**	0,257	**0,829**	0,489	**0,935**	-0,141	**0,620**	0,489	0,501	**0,918**	**0,841**	1

3.3.2. Correlation Analysis

High positive correlations were found among the cations (r = 0.92 between Ca and K, Na and K was r = 0.98, between Ca and Na, r = 0.93). Among the anions, high positive correlations exist between NO_3^- and Cl^-, r = 0.69. Since the correlation coefficient between the cation and anion pairs is positively high, it can also be deduced that most of the water samples (omitting the rainwater samples) originate from a common source. Very high positive correlations exist between total dissolved solids (TDS) and Ca, Na, K, HCO_3 and Cl^-, respectively r = 0.94, 0.84, 0.83, 0.92 and 0.62. These correlations between ions and TDS suggest that TDS is derived mainly from Ca, Na, and HCO_3^-. Other moderate correlations were observed between NH_4^+ and Mg^{2+}, Ca^{2+} and Cl^-, Ca^{2+} and Mg^{2+}, NH_4^+ and SO_4^{2-} and between K^+ and NO_3^-. Among the ions, Ca^{2+} makes the highest contribution followed by HCO_3^-, Cl^- and SO_4^{2-} indicating the incorporation of soil dust into the rain samples, which reflects a major crustal influence (Al-Khashman, 2009).

It is quite clear that most of the parameters were significantly correlated with HCO_3^-, which could be an indication that the aquifer system may have experienced various processes such as ion exchange, water-rock interaction and weathering of the aquifer's parent material, therefore making HCO_3^- a dominant ion in the water chemistry (Kura et al., 2013).

For instance, the correlation matrix shows high positive correlation between Ca and HCO_3^-, which indicates recharge; at the same time it can also be an indication of weathering of calcite mineral, as illustrated in Equations (8) (carbonic acid formation) and (9) (calcite weathering equation):

$$H_2O + CO_2 \rightarrow H_2CO_3 \qquad (8)$$

$$CaCO_3 + CO_2 + H_2O \rightarrow Ca^{2+} + 2HCO_3^- \qquad (9)$$

The variations of Ca^{2+}, Na^+, and HCO_3^- were derived from natural weathering reactions, which suggested that the same geochemical processes occurred in the rock weathering reactions that released Ca^{2+}, Na^+, and HCO_3^- into the groundwater along the flow path. This interpretation is consistent with the water-rock interaction equation for basaltic aquifers (Zhang et al., 2012; Lu et al., 2015).

Based on the strong correlation between NO_3^- and Cl^- (r = 0.69) and the absence of a significant correlation between NO_3^- and K^+ (r = 0.41), this suggests an anthropogenic source of NO_3^- from the use of agrochemicals in the surrounding plantations and inputs from domestic waste (sewage sludge, effluent) that infiltrates into the groundwater sources in the study area (Ako et al., 2011; Wirmvem et al., 2013; Wotany et al., 2013; (Gountié, et al., 2017)).

3.4. Hydrochemistry

3.4.1. Hydrochemical Facies

The chemical composition of the water samples was identified and classified using the trilinear plot described by Piper (1944), and the data are presented on Figure 4.

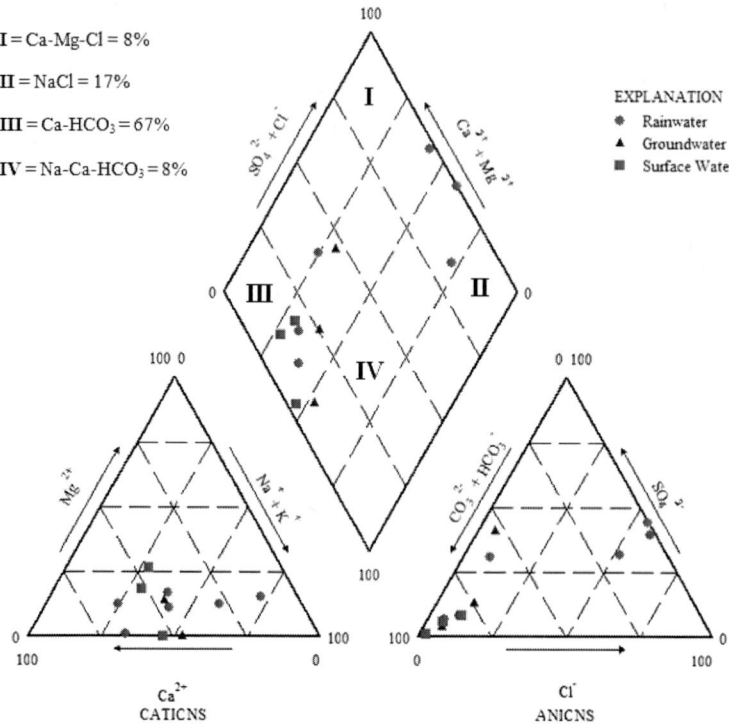

Figure 4. Piper diagram showing water types in the study area.

From the Piper's diagram above, following the defined chemical zones it shows that eight of the twelve samples (67%) have a Ca-HCO$_3$ domain, two samples (17%) are of the NaCl domain and one other from the Na-Ca-HCO$_3$ and Ca-Mg-Cl domains each. This plot shows that the samples are chemically dominated by HCO$_3^-$ and Cl$^-$ anions and Ca^{2+} and to some extent Na$^+$ were the major cationic species.

3.4.2. Mechanisms Controlling Water Chemistry

From the Piper diagram, most samples plotted in the field of Ca+Na+K. This indicates their dominance in the groundwater aquifer system. The observed spatial distribution of the water types showed no discernible pattern from Ca-HCO$_3$ to Na-Cl. The major ion compositions varied in the groundwater samples. However, these samples were generally HCO_3^--rich with no dominant type of cation (Figure 4).

The varieties of specific water types of one place or region increase accordingly with the variety of geology and climate. On young volcanic watersheds (Ca-HCO_3 water type) and also on less extended old crystallinic ones (Na-Cl water type) occurs (Weninger, 1985).

The predominance of HCO_3^- suggests that intense chemical weathering processes are taking place in this aquifer. Natural processes such as the diverse dissolution of silicate minerals in the rocks that react with CO_2 gas from root or microorganism respiration and from the mineralization of soil organic matter could be a mechanism that releases Ca, Mg, Na, K and HCO_3 into the groundwater, as indicated in the following reactions:

Anorthite: $CaAl_2Si_2O_8 + 2CO_2 + 3H_2O \rightarrow Al_2Si_2O_5(OH)_4 + Ca^{2+} + 2HCO_3^-$ (10)

Albite: $2NaAlSi_3O_8 + 9H_2O + 2H^+ \rightarrow Al_2Si_2O_5(OH)_4 + 4H_4SiO_4 + 2Na^+$ (11)

Amphibole: $Ca_2Mg_5Si_8O_{22}(OH)_2 + 14CO_2 + 22H_2O \rightarrow 2Ca^{2+} + 5Mg^{2+} + 14HCO_3^- + 8Si(OH)_4$ (12)

Pyroxene: $CaMg(Si_2O_6) + 4CO_2 + 6H_2O \rightarrow Ca^{2+} + Mg^{2+} + 4HCO_3^- + 2Si(OH)_4$ (13)

Orthoclase: $2KAlSi_3O_8 + 11H_2O \rightarrow Si_2O_5Al_2(OH)_4 + Si(OH)_4 + 2K^+ + 2OH^-$ (14)

Similar observations have been reported in springs along the CVL (Tanyileke et al., 1996, Temgoua et al., 2008, Ako et al., 2012) and elsewhere in other countries. This confirms the dominance of HCO_3^- in the groundwater sources as a result of weathering reactions on rocks in the study area by percolating groundwater. Where silicate mineral weathering is the major controlling process, concentrations of the major physico-chemical parameters are relatively low as with the case of the study area (Yidana et al., 2010).

The action of microorganisms found in soils also plays a significant role in the formation of bicarbonate ions through the breakdown of vegetable matter according to the following equation (Ako et al., 2011; Gountié et al., 2017):

$$H_2CO_3 \rightleftharpoons HCO_3^- + H^+ \rightleftharpoons CO_3^{2-} + H^+ \qquad (15)$$

3.4.3. Hydrochemical Evolution

The major ion chemistry of groundwater is an important tool for determining solute sources and for describing groundwater evolution as a result of water–rock interaction leading to the dissolution of carbonate minerals, silicate weathering and ion exchange processes (Kumar et al., 2007; Aghazadeh et al., 2017). Results from the chemical analyses were used to identify the geochemical processes and mechanisms in the groundwater aquifer system.

The data of water samples were plotted on a Gibbs's diagrams (Gibbs, 1970, Figure 5a,b). The clustering of the data points in the Gibbs diagram indicates that chemical weathering of rock-forming minerals is the main factor controlling the water chemistry in the study area. Thus, the interaction between rocks and water results in leaching of ions into the groundwater system which influences the water chemistry.

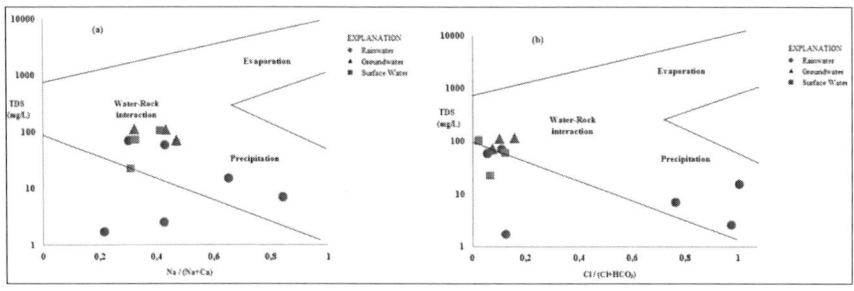

Figure 5. Gibbs (1970) plots indicating water-rock interaction as the main process regulating the chemistry of waters in the study area. (a) TDS versus Na/(Na+Ca) diagram; (b) TDS versus Cl/(Cl+HCO3).

The Gibbs diagram for the study area (Figure 5a, b) which indicates that water-rock interaction is the dominant process controlling the main chemical composition of the groundwater and surface water sampled.

To determine the type and the degree of rock weathering in the study area, the weathering index (RE) was used (Tardy, 1971).

The weathering index, RE can be used to characterize water–rock interaction in the study area. The calculation of this index is written as (Negrel, 1999):

$$RE = 2(3K + 3Na + 2Ca - Si) / (K + Na + Ca) \qquad (16)$$

All elements are in mol.

According to Tardy (1971), this ratio is equivalent to the $(SiO_2)/(Al_2O_3)$ molar ratio of the secondary mineral newly formed in soil profiles. A scale of RE variations corresponding to the type of weathering shows that if:

RE < 0, the weathering process is allitization and gibbsite is the main mineral formed.

0 < RE < 2, the weathering process is monosiallitization, which results in kaolinite formation.

RE > 2, the weathering process is bisiallitization then kaolinite and montmorillonite (smectites) are the main minerals formed.

RE for Mt. Cameroon water samples vary between 4.96 and 5.28 (mean, 5.09) with all the water samples having RE > 4 (Table 1). This suggests that bisialitization is the main weathering processes taking place in the study area. Under this weathering process, weak to moderate leaching favors an approximate balance between SiO_2 and cations which yields the formation of 2:1 clays (Tardy, 1971). This result is in accordance with those obtained from the Gibbs plot approach.

Also, the finding is consistent with those reported in springs of the Mt. Cameroon Area, in the Lake Nyos catchment, in the Lake Monoun area by Ako et al., 2012; Fantong et al. 2015; Kamtchueng et al., 2016 respectively using the chemical composition of the granitic bedrock (monzonite) and basaltic deposits which are similar lithologies to the study area.

The Mt. Cameroon area is bounded on its western borders by the Atlantic Ocean and this closeness to the sea might affect the water chemistry and quality of rain water falling in the area.

The most usual method of evaluating the contribution of sea salts to ion contents in rainwater is to compare the Cl^-/Na^+ ratio in the rainwater to that of seawater. The sea is considered to be the major source of both ions, although they may also be emitted from other natural and anthropogenic sources (Wang and Han, 2011). In areas close to the sea, sea salt is a major contributor to sodium and chloride deposition, and the Cl/Na ratio in precipitation is typically equal to that of sea salt. Hence, calculation of the Cl/Na ratio is useful in sites that are far from the sea, with a value below 1.17 considered to be acceptable (Berner and Berner, 1987).

The contribution of marine component for a given element X (X=Cl^-, NO_3^-, SO_4^{2-}, K^+, Ca^{2+} and Mg^{2+}) is given by:

$$EF = \frac{(X/Na^+)rain}{(X/Na^+)seawater} \quad (17)$$

Where X is the ion of interest.

The enrichment factors (EF) for ions (X) have been calculated using Na^+ as the marine reference and are given in Table 3 below.

Table 3. Ratio of equivalent concentration of ions with reference to Na+

	SO_4^{2-}/Na^+	NO_3^-/Na^+	Cl^-/Na^+	Ca^{2+}/Na^+	Mg^{2+}/Na^+	K^+/Na^+
Sea water [a]	0.12	0.00002	1.17	0.044	0.23	0.022
Water Samples	1.72	1.58	1.25	1.65	0.44	0.24
Enrichment Factor	14.30	78763.78	1.07	37.60	1.91	10.94

[a] Berner and Berner (1987)

The elemental ratios (X/Na) were determined according to the composition of seawater given by Berner and Berner (1987).

The mean value for Cl^-/Na^+ in the water samples is 1.25. This value is approximately equal to that of 1.17 in seawater, indicating that almost all

Na$^+$ and Cl$^-$ in water samples is derived from the sea. The SO$_4^{2-}$/Na$^+$ ratio of 1.72 in the water samples indicated that there is almost no marine contribution of SO$_4^{2-}$. Charcoal-fired fish roasting activities and traffic emission could account for the enrichment of sulfate in rain in the study area. Great enrichment of NO$_3^-$ was attributed to the traffic emission and the use of fertilizers in agricultural activities.

The marine contribution to Ca^{2+} and K$^+$ was also negligible based on the high EF of Ca^{2+} and K$^+$ (Table 3). On the other hand, Mg^{2+} had a mean value of 0.44 which is close to the seawater levels for Mg^{2+} (Mg$_{seawater}$ = 0.23), this indicates that some of the Mg^{2+} could be of marine origin.

These results indicate that the water samples were influenced by seawater in the study region. Therefore, we can conclude that wind-blown salts are carried from the sea and recharge groundwater through rainfall. In conclusion, their groundwater chemistry was potentially dominated by rock weathering and seawater (Lu et al., 2015). From Table 3, it can be noted that the EF values are high for NO$_3^-$, SO$_4^{2-}$, Ca^{2+}, and K$^+$ whereas Cl$^-$ and Mg^{2+} had values similar to the seawater standards.

CONCLUSION

This study is essential for establishing a data base about the rainwater quality around one of the most important watersheds Cameroon.

The study of water in the area revealed that the samples were generally weakly acidic -alkaline (4.24 < pH < 8), fresh (TDS < 110 mg/L) and slightly mineralized (EC < 370 µS/cm). Major ion concentrations were low, and below maximum values of the WHO standards for drinking water. The relative abundance of major cations and anions in the rainwater (mg/L) is Ca^{2+}>Na$^+$>Mg^{2+}>K$^+$>NH$_4^+$ and HCO$_3^-$>NO$_3^-$>Cl$^-$>SO$_4^{2-}$>F$^-$, respectively.

The main water types and proportions are Ca-HCO$_3$ type (67%) and Na-Cl type (17%). This shows that the samples are chemically dominated by HCO$_3^-$ and Cl$^-$ anions however Ca^{2+} and to some extent Na$^+$ were the major cationic species. The chemistry of the water showed that a variety of natural processes and anthropogenic activities are responsible for the diversity of the hydrochemical facies. These natural processes are dissolution, ion exchange capacity from water-rock interactions; these reactions are governed in most cases by a CO_2 gas phase of biogenic origin, whereas soil dust, agriculture and biomass burning account for the anthropogenic sources. The weathering type, characterized from the geochemical ratio RE between dissolved cations in the Mt. Cameroon sampled waters, corresponds to bisiallitization indicating the genesis of smectites where kaolinite and montmorillonite are the main minerals formed.

REFERENCES

Adomako, D., Osae, S., Akiti, T. T., Faye, S., Maloszewski, P. (2011). Geochemical and isotopic studies of groundwater conditions in the Densu River Basin of Ghana. *Environmental Earth Sciences,* 62: 1071-1084.

Aghazadeh, N., Chitsazan, M., Golestan, Y. (2017). Hydrochemistry and quality assessment of groundwater in the Ardabil area, Iran. *Applied Water Science,* 7: 3599-3616.

Ako, A. A., Shimada, J., Hosono, T., Kagabu, M., Akoachere, R., Nkeng, G. L., Eneke, T. E. G., Fouepe, A. L. (2012). Spring water quality and usability in the Mount Cameroon area revealed by hydrogeochemistry. *Environmental Geochemistry and Health,* 34: 615-639.

Ako, A. A., Shimada, J., Hosono, T., Ichiyanagi, K., Nkeng, G. E., Fantong, W. Y., Eneke, T. E. G, Njila, N. R. (2011). Evaluation of groundwater quality and its suitability for drinking, domestic, and agricultural uses in the Banana Plain (Mbanga, Njombe, Penja) of the Cameroon Volcanic Line. *Environmental Geochemistry and Health,* 33: 559-575.

Al-Khashman, O. A. (2009). Chemical characteristics of rainwater collected at a western site of Jordan. *Atmospheric Research,* 91: 53-61.

Appelo, C. A. J., Postma, D. (2005). *Geochemistry, Groundwater, and Pollution*, 2nd ed.; Balkema Publishers: Rotterdam, The Netherlands.

Berner, E. K., Berner, R. A. (1987). *The global water cycle: geochemistry and environment*, Englewood Cliffs: Prentice-Hall.

Djieto Lordon, A. E., Agyingi, C. M., Manga, V. E., Bukalo, N. N., Beka, E. T. (2017). Geo-electrical and borehole investigation of groundwater in some basalts on the South-Eastern Flank of Mount Cameroon. *West Africa Journal of Water Resource and Protection,* 9: 1526-1546.

Edet, A., Nganje, T. N., Ukpong, A. L., Ekwere, A. S. (2011). Groundwater chemistry and quality of Nigeria: A status review. *African Journal of Environmental Science and Technology,* 5 (13): 1152-1169.

Edmunds, W. M., Smedley, P. L. (1996). Groundwater geochemistry and health: An overview. In: Appleton, J. D., Fuge, R., McCall, G. J. H. (eds.) Environmental geochemistry and health with special reference to developing countries. *Geological Society Special Publication,* 113: 91-105.

Fantong, W. Y., Satake, H., Ayonghe, S. N., Aka, F. T., Asai, K. (2009). Hydrogeochemical controls and usability of groundwater in semi-arid Mayo Tsanaga River Basin, Far-North Cameroon. *Environmental Geology,* 58 (6): 1281-1293.

Fantong, W. Y., Kamtchueng, B. T., Yamaguchi, K., Ueda, A., Issa, Ntchantcho, R., Wirmvem, M. J., Kusakabe, M., Ohba, T., Zhang, J., Aka, F. T., Tanyileke, G., Hell, J. V. (2015). Characteristics of chemical weathering and water-rock interaction in Lake Nyos dam (Cameroon): implications for vulnerability to failure and re-enforcement. *Journal of African Earth Science,* 101:42-55.

Freeze, R. A., Cherry, J. A., (1979). *Groundwater;* Prentice Hall: Englewood Cliffs, NJ, USA, pp. 604.

Gibbs, R. J. (1970). Mechanisms controlling world water chemistry. *Science,* 17: 1088-1090.

Gnazou, M. D. T., Bawa, L. M., Banton, O., Djanéyé-boundjou, G. (2011). Hydrogeochemical characterization of the coastal Paleocene aquifer of Togo (West Africa). *International Journal of Water Resources and Environmental Engineering*, 3(1): 10-29.

Gountié, M. D., Tsozué, D., Mimba, M. E., Fulbert, T., Nembungwe, R. M., Linida, S. (2017). Importance of rocks and their weathering products on groundwater quality in Central-East Cameroon. *Hydrology*, 4 (23): doi:10.3390/hydrology4020023.

Hem, J. D. (1989). *Study and interpretation of the chemical characteristics of natural water*. U.S Geological Survey Water-Supply Paper; U.S Geological Survey: Reston, VA, USA, pp. 272.

Kamtchueng, B. T., Fantong, W. Y., Wirmvem, M. J., Tiodjio, R. E., Takounjou, A. F., Ndam Ngoupayou, J. R., Kusakabe, M., Zhang, J., Ohba, T., Tanyileke, G. (2016). Hydrogeochemistry and quality of surface water and groundwater in the vicinity of Lake Monoun, West Cameroon: Approach from multivariate statistical analysis and stable isotopic characterization. *Environmental Monitoring and Assessment*, 524:1-24.

Karklins, S. (1996). *Groundwater sampling desk reference*. Wisconsin Department of Natural Resources Bureau of Drinking Water and Groundwater, PUBL-DG-037 96.

Kulshrestha, U., Monika, T., Kulshrestha, M., Sekar, R., Vairamani, M. (2003). Chemical characteristics of rainwater at an urban site of South Central India, *Atmospheric Environment*, 37:3019-3026.

Kumar, M., Kumari, K., Ramanathan, A. I., Saxena, R. (2007). A comparative evaluation of groundwater suitability for irrigation and drinking purposes in two intensively cultivated districts of Punjab, India. *Environmental Geology*, 53: 553-574.

Kura, N. U., Ramli, M. F., Sulaiman, W. N. A., Ibrahim, S., Aris, A. Z., Mustapha, A. (2013). Evaluation of factors influencing the groundwater chemistry in a small tropical island of Malaysia. *International Journal of Environmental. Research and Public Health*, 10: 1861-1881; doi:10.3390/ijerph10051861.

Liotta, M., Brusca, L., Grassa, F., Inguaggiato, S., Longo, M., Madonia, P. (2006). Geochemistry of rainfall at Stromboli volcano (Aeolian Islands): Isotopic composition and plume-rain interaction. *Geochemistry Geophysics Geosystems,* **7**, Q07006. doi:10.1029/2006 GC001288.

Lu; Y., Tang, C., Chen, J. (2015). Groundwater recharge and hydrogeochemical evolution in Leizhou Peninsula, China. Hindawi Publishing Corporation *Journal of Chemistry,* Article ID 427579, 12 pages http://dx.doi.org/10.1155/2015/427579.

Madonia, P., Liotta, M. (2010). Chemical composition of precipitation at Mt. Vesuvius and Vulcano Island, Italy: volcanological and environmental implications. *Environmental Earth Science,* 61: 159-171.

Magha, A., Awah, M. A., Nono, G. D. K., Wotchoko, P., Tabot, M. A., Kabeyene, V. K. (2015). Physico-chemical and bacteriological characterization of spring and well water in Bamenda III (NW Region, Cameroon). *American Journal of Environmental Protection,* 4 (3): 163-173.

Mather, T. A., Oppenheimer, C., Allen, A. G., McGonigle, A. J. S. (2004). Aerosol chemistry of emissions from three contrasting volcanoes in Italy. *Atmospheric Environment,* 38 (33):5637-5649. doi:10.1016/ j.atmosenv.2004.06.017.

Mohapatra, P., Vijay, R., Pujari, P., Sundaray, S., Mohanty, B. (2011). Determination of processes affecting groundwater quality in the coastal aquifer beneath Puri city, India: A multivariate statistical approach. *Water Science and Technology,* 64: 809-817.

Morgenstern, U., Brown, L. J., Begg, J., Daughney, C., Davidson, P. (2009). Linkwater Catchment groundwater residence time, flow pattern, and hydrochemistry trends. *Gns Science Report* 2009/08.

Nduka, J. K., Orisakwe, O. E. (2011). Water-quality issues in the Niger Delta of Nigeria: A look at heavy metal levels and some physicochemical properties. *Environmental Science and Pollution. Research,* 18: 237-246.

Negrel, P. (1999). Geochemical study of a granitic area - the Margeride Mountains, France: Chemical element behavior and $^{87}Sr/^{86}Sr$ constraints. *Aquatic Geochemistry,* 5:125-165.

Njitchoua, R., Ngounou-Ngatcha, B. (1997). Hydrogeochemistry and environmental isotope investigations of the North Diamare Plain Northern Cameroon. *Journal of Africa Earth Science,* 25 (2): 307-316.

Oppenheimer, C. (2003). Volcanic degassing. In: Holland, H. D., Turekian, K. K. (eds.) *The crust, treatise on geochemistry.* Elsevier-Pergamon, Oxford, 3:123-166.

Piper, A. M. (1944). A graphical interpretation of water-analysis. *Transaction of the American Geophysical Union,* 25:914-928.

Tanyileke, G., Kusakabe, M., Evans, W. C. (1996). Chemical and isotopic characteristics of fluids along the Cameroon Volcanic Line, Cameroon. *Journal of Africa Earth Science,* 22: 433-441.

Tardy, Y. (1971). Characterization of the principal weathering types by the geochemistry of waters from some European and African crystalline massifs. *Chemical Geology,* 7: 253-271.

Temgoua, E., Djeuda, T. H. B., Tanawa, E., Guenat, C., Pfeifer, H. R. (2005). Groundwater fluctuations and footslope ferricrete soils in the humid tropical zone of southern Cameroon. *Hydrological Processes,* 19: 3097-3111.

Wang, H., Han, G. L. (2011). Chemical composition of rainwater and anthropogenic influences in Chengdu, Southwest China, *Atmospheric Research,* 99 (2): 190-196.

WHO. (2008). *Guidelines for drinking-water quality,* Third Edition, Incorporating the First and Second Addenda, Volume 1: Recommendations, WHO, Geneva.

William, A. B., Benson, N. U. (2010). Interseasonal hydrological characteristics and variabilities in surface water of tropical estuarine ecosystems within Niger Delta, Nigeria. *Environmental Monitoring and Assessment,* 165: 399-406.

Wirmvem, J. M., Ohba, T., Fantong, W. Y., Ayonghe, S. N., Suila, J. Y., Asaah, A. N. E., Tanyileke, G., Hell, J. V. (2013). Hydrochemistry of shallow groundwater and surface water in the Ndop plain, North West

Cameroon. *African Journal of Environmental Science and Technology,* 7(6):518-530.

Wotany, E. R., Ayonghe, S. N., Fantong, W. Y., Wirmvem, M. J., Ohba, T. (2013). Hydrogeochemical and anthropogenic influence on the quality of water sources in the Rio del Rey Basin, South Western, Cameroon, Gulf of Guinea. *African Journal of Environmental Science and Technology,* 7(12): 1053-1069, doi: 10.5897/AJEST2013.1578.

Yidana, S. M. (2010). Groundwater classification using multivariate statistical methods: Southern Ghana. *Journal of Africa Earth Science,* 57, pp. 455-469.

Zhang, B., Song, X., Zhang, Y. (2012). Hydrochemical characteristics and water quality assessment of surface water and groundwater in Songnen plain, Northeast China, *Water Research,* 46 (8): 2737-2748.

In: Groundwater Quality
Editor: Rafael M. Vick

ISBN: 978-1-53618-807-3
© 2020 Nova Science Publishers, Inc.

Chapter 4

HIGH EXPOSURE DOSE OF FLUORIDE OWING TO RISK OF FLUOROSIS IN INHABITANTS OF SARENI BLOCK LOCATED AT THE GANGA RIVER BASIN UTTAR PRADESH (INDIA)

Pokhraj Sahu[1], PhD, Pramod Kumar Singh[2], PhD, Prashant Singh[3], PhD, Vinay Kumar[2], PhD and Pramod Kumar[4], MSc

[1]Chemical Laboratory, Bharat Oil and Waste Management Ltd., Roorkee, (India)
[2]Department of Chemistry, Babu Banarasi Das University, Lucknow (India)
[3]Central Pollution Control Board (North Zone), Lucknow, India
[4]Environmental Monitoring Laboratory, CSIR-Indian Institute of Toxicology Research, Lucknow (India)

Abstract

In India, during 1991, an estimate of 66 million was at risk of fluorosis which reached up to 120 million in 2018, it was an alarming situation and it seemed necessary to take some vital steps in order to address the fatal and incurable problem in the country. A study was conducted to assess the extent of exposure dose of fluoride in inhabitants of Sareni block of Raebareli district, Uttar Pradesh, where in, twenty-five groundwater samples were collected from different places in different direction (center, north, south, east and west zone) of Sareni block. The mean concentration of fluoride was found to be 1.19, 1.27, 1.23, 1.52, and 1.19 mg/L in center, north, south, east and west zones respectively, which is higher than Indian drinking water standard. The estimated dose of exposure of fluoride in the drinking water, as per the investigation in the drinking, were ~8, ~4 and ~ 2 times higher than the intake dose of fluoride recommended by Agency for Toxic Substances and Disease Registry for infants, children, and adults respectively. The results indicated that children were highly closed to the risk of incurable fluorosis as compare to adults in the study area because of high exposure and absorption efficiency of fluoride. Water quality assessment is one of the most important steps of water management practice for it is safe in drinking water uses or other specific purpose. This study will generates baseline data about the fluoride contaminated area of Sareni block.

Keywords: exposure dose, fluoride, fluorosis, groundwater, risk

Introduction

Incurable health hazard in human and other animals are increasing day by day across the globe due to intake of potable water contaminated with fluoride. In India, groundwater is chiefly using as a drinking watersources, which is naturally contaminated with fluoride, arsenic, nitrate, sulphate, and heavy metals but, contamination with fluoride is more spreading as compared to arsenic (Muralidharan et al., 2002). Fluoride is an essential element, which is distributed in wide range all over the world.The elevated concentration of fluoride in groundwater was reported in 335 districts of India. Fluoride is a natural constituents and 13th most abundant element in the earth's crust. The beneficial level of fluoride is ranged from 0.5 to 1.0 mg/L for the development of bones and teeth of human, below and above the level can cause significant health hazards (Thivya et al., 2014). Fluoride, below the acceptable recommended level (0.5 mg/L) in potable water can cause dental caries while beyond the permissible level(1.5 mg/L) may lead to dental fluorosis (Mondal et al., 2012). In tropical or hot climate zone, the drinking water having concentration between 1.0 to 1.5 mg/L can cause to fluorosis because of the high rate of water consumption can increase the ingested quantity of fluoride in the body. The long-term intake of an excessive level (1.5 to 4 mg/L) of fluoride containing portable water can magnify the normal mineralization process to hypo-mineralization in opaque white tooth enamel, and this opaque enamel is converted into pitting and molted of teeth. Ingestion of fluoride-containing drinking water beyond 4 mg/L can cause to skeletal fluorosis while exceeding 10 mg/L causes to crippling fluorosis and non-skeletal fluorosis (Kundu et al., 2011). The Skeletal fluorosis is calcification of bones, joint, and ligaments due to chronic consumption of excessive fluoride that creating pain in bones, and joint. Non- skeletal fluorosis such as vomiting, nausea, muscle fibre degeneration, neurological effect, low haemoglobin levels, skin rashes, abdominal pain, gastrointestinal problems, destruction of sixty enzymes etc. (Meenakshi and Maheshwari 2006). Ignition of fluoride through dietary sources plays a secondary role in the development of fluorosis.

An elevated level of fluoride was found in many parts of the world along with that there are 85 million tons of fluoride deposits in the earth crust. Two hundred million people are affected in 27 countries in which the three most affected nations are Africa, China, and India (Bhattacharya and Chakrabarti 2011). Twelve million tons of fluoride deposited in the Indian continental crust and 120 million peoples are prone to the high risk of fluorosis in over 335 districts in 20 states of India. Fluoride in groundwater was earliest identified in 1937 in Nellore district of Andhra Pradesh. Gujarat, Rajasthan, Andhra Pradesh, Bihar, Tamil Nadu, Madhya Pradesh, Uttar Pradesh are worst affected states in India. Prakasam district of Andhra Pradesh is foremost affected district in India, where the ground water of 195 villages has been contaminated with fluoride (Subba Rao et al., 2017).

Fluoride contamination in the water is the result of both natural and manmade activity. Manmade activity such as application and manufacturing of phosphate fertilizer and pesticides, sewage and sludge used in agriculture practice, coal-based thermal power plants, coal-based bricks manufacturing, hydrofluoric acid and sodium fluoride manufacturing plant, manufacturing of integrated circuit and semiconductor are different sources of fluoride contamination in water bodies. The volcanic eruption and geological formation are natural causes of fluoride contamination in groundwater. The natural or geogenic origin is the major sources of fluoride contamination of groundwater of India. The fluoride-containing minerals are present in the geology of the contaminated area which is mainly responsible of raising the level of fluoride in water. Fluoride bearing minerals are apatite, hornblende, fluorite, and biotite etc. (Sahu, et al., 2019; Subba Rao et al., 2017). The dissolution of fluoride depends on the solubility of fluoride bearing minerals, geological formation, groundwater flow direction, an ions-exchange capacity of aquifer materials, the climate of a region and water quality parameter such as pH, bicarbonate, temperature, sodium etc. (Saxena and Saxena 2014). Fluoride concentration went beyond the limits in many districts of Uttar Pradesh such as Firozabad, Agra, Pratapgarh, Aligarh, Mathura, Raebareli, Jaunpur, Unnao, Fatehpur, Kannauj, Varanasi, Mau, and Jhansi (Sahu et al., 2018; Saxena and Saxena 2014; Raju et al., 2012; Chandha et al., 1999; Gupta et al., 1999). The exposure dose/chronic daily intake of fluoride through potable water as estimated early in many parts of India (Shukla and Saxena 2020; Karunanidhi et al., 2019), but it was laking for Sareni block.

The key objective of the present study is to evaluate the fluoride contents in groundwater and examine exposure dose to predict and analyze the associated health risk in inhabitants of Sareni block. This study might generate baseline record which will be used for planning and water resource management.

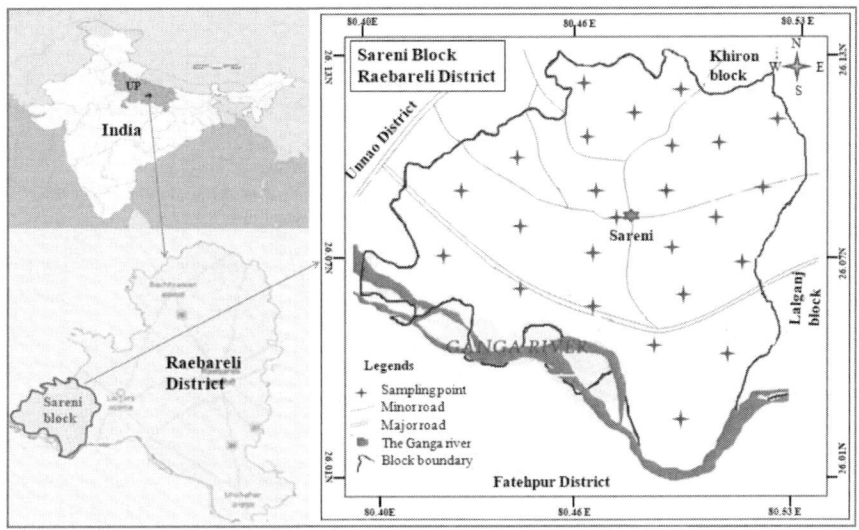

Figure 1. Demographical detail of groundwater sampling location of Sareni block.

MATERIALS AND METHODS

Study Area

The Sareni block lies between latitude 26°01' N and 26° 13' N and longitude 80° 40' E and 80° 53' E shown in Figure 1, covering a 277.57 square kilometer area in south-west part of the Raebareli district. The Sareni block situated at the Ganga river basin and the population of this block was 1, 79,687 including 20,935 population of 0 to 6-year children as per Census 2011. Sareni block is situated in Raebareli district and it 6.02% land area and 5.28% of population of Raebareli district. The district has a subtropical climate i.e., it is hot in summer from April to June (40 to 45°C) and very cold and dry in winter from December to February (3 to 12 °C). The fall season started from Mid-June to September coming from south-west monsoon. The major part of the district is devoid of any hard rocks and alluvium deposition of quaternary age. Sodic soil is spread out over the district with soil texture of that of loamy sand, clay loam, silt loam, and sandy loam, respectively. According to, the fluoride in groundwater governs by muscovite and biotite fluoride minerals as in clay, and mud around the Indo-Gangetic basin (Kumar and Saxena 2011). According to CGWB, aquifer of the district found a 4 stage aquifer system has been recognized down up to a depth of 600 m bgl (below ground level). According to Central Ground Water Board, the groundwater resource of Sareni block was categorized under the critical category and water level of contour zone of Sareni block found between 5 to 10 m bgl in may, 2014.

Sample Collection and Analysis

The total twenty-five groundwater samples were collected from bore well and Hand pumps of Sareni block of Raebareli district, Uttar Pradesh (India) during pre-monsoon 2016. Five-five samples were drawn from each direction (North, south, east and west) with center location also taken around the Sareni village for the collection of homogenized groundwater samples all over the blocks. All samples were collected in polypropylene plastic sampling bottle after five minute pumping of bore well and hand pumps. Eight physicochemical water quality parameters were analyzed. Calcium, sodium, and magnesium were analyzed by following flame emission method through atomic absorption spectrophotometer, and total dissolve solid was estimated by gravimetric method. Total hardness and alkalinity or bicarbonate was examined by the titration method. The pH and fluoride were analyzed by ion selective electrode (Orion 4 Star) approved by environmental protection agency for the measurements of fluoride in water and wastewater {standard method 4500-F-C and ASTM D1179 (B)}.

Statistical Analysis

A linear correlation was applied between fluoride and pH, TDS, TA, TH, Ca^{2+}, Na^+ in sigma stat software, while cluster analysis was performed in SPSS software by ward linkage method base on fluoride concentration in all part of Sareni block.

Exposure Dose Assessment of Fluoride and Water Intake

Estimation of intake of fluoride through drinking water in human population was classified in infant, children and adult their body weight ≤6 kg, 7 kg to 20 kg and 20 kgto 70 kg respectively (Jha et al., 2009). Average intake for infant, children, and adults were investigated 1, 3 and 6 Liter respectively. The exposure dose of Fluoride was calculated by following formula for different age groups.

$$ED = (C/BW) * WI \quad\quad\quad\quad\quad\quad\quad\quad\quad\quad \text{(Eq. 1)}$$

where ED is fluoride exposure dose (mg kg^{-1} day^{-1}), C is fluoride content in water (mg/L), BW is body weight (kg), and WI is water intake (L/day).

Quality Control and Quality Assurance

NABL certified glass ware and AR grade chemical were used for analysis. All glassware was dipped overnight into 0.15 N nitric acid solutions and washed with double distilled water before using. Before analysis of samples ISE was calibrated by the standard working solution of NaF (10 ml of 100, 10, 1 and 0.1 ppm F$^-$ solution) with the pouring of 1 mL of TISHAB-III (total ionic strength buffer solution) solution. After that fluoride was measured by 10 ml of water samples were taken in a plastic beaker with 1 mL of TISHAB-III. After measurement of five sample fluoride meter was cross-checked by the standard solution.

Results and Discussion

Evaluation for Drinking Water Quality

Groundwater is precious and a core source of fresh water, but now a day's excessive application of fertilizers and pesticides on agricultural land contaminates surface water and it percolating to aquifer, is an example of anthropogenic groundwater pollution, while natural groundwater pollution is arising from the chemical composition of geology and over exploitation of groundwater can contribute to 80% human disease and 30% mortality in an infant of developing countries (Chakraborty 1999). The groundwater of Sareni block was alkaline in nature and pH of centre, north, south, east and west zone ware ranged from 7.3 to 8.1, 7.0 to 7.62, 7.39 to 8.21, 7.01 to 7.62 and 7.12 to 7.50 respectively. The pH of all zones was under the prescribed limit of Indian and WHO standard given in table 1. All dissolved minerals contributes to total dissolved solids are an important parameter of drinking water quality because we need to minerals for growth and hormonal activation. The average TDS were found 821, 1222, 752, 1284 and 707 mg/L in centre, north, south, east and west zone respectively of Sareni block. The TDS values were shown above the desirable limits (500 mg/L) but below the permissible limits (2000 mg/L) of drinking water criteria. 80% groundwater samples of north and east zone of Sareni block were found above the desirable limits can causes for concern given in table 1. Consumption of high TDS containing drinking water can cause gastro-irritation to consumer. Total alkalinity as neutralized by the capacity of acid, was found beyond the desirable limits but under permissible limits and range from 254 to 725 mg/L. The minimum concentration of total alkalinity found in west zone and maximum was found in the north zone. The average value of total hardness was found in 242, 428, 241, 448, and 226 mg/L for the centre, north, south, east and west zone of Sareni block, which was below the permissible limits but beyond the desirable limits. Total alkalinity and total hardness both are contributing foremost portion for total dissolved solids.

Table 1. Results of physico-chemical facies in groundwater in different direction of Sareni block

Statistical	pH	TDS	TA	TH	F⁻	Ca^{2+}	Mg^{2+}	Na^+
Center Zone								
Mean	7.58	821	319	242	1.19	33	66	163
Min	7.30	722	288	204	0.80	13	53	132
Max	8.01	985	380	301	1.84	45	83	212
SEM	0.12	46.23	16.96	16.09	0.19	5.54	5.34	14.40
North Zone								
Mean	7.42	1222	409	428	1.27	61	95	236
Min	7.00	604	299	232	0.64	22	54	120
Max	7.62	2053	725	688	2.21	133	151	321
SEM	0.11	242.69	79.50	79.15	0.28	19.20	16.04	32.69
South Zone								
Mean	7.65	752	383	241	1.23	43	53	135
Min	7.39	617	298	222	0.64	39	42	111
Max	8.21	905	526	270	2.12	53	67	169
SEM	0.15	45.74	42.04	8.03	0.25	2.61	4.04	9.95
East Zone								
Mean	7.26	1284	431	448	1.52	53	126	178
Min	7.01	991	352	356	1.02	39	93	165
Max	7.62	1813	523	652	2.00	62	200	189
SEM	0.10	144.68	27.26	54.89	0.19	4.23	19.89	4.09
West Zone								
Mean	7.30	707	362	226	1.19	40	50	149
Min	7.12	547	254	196	0.94	22	24	134
Max	7.50	903	484	286	1.88	57	73	180
SEM	0.07	62.83	39.68	16.32	0.17	5.77	8.30	8.38

Table 1. (Continued)

Statistical	pH	TDS	TA	TH	F-	Ca2+	Mg2	Na⁺
Indian Standard 10:500:2012 (b)	(a) 6.5	500	200	200	1.0	75	30	-
	8.5	1000	600	600	1.5	250	150	-
WHO 2011 (a)	6.5	500	-	-	-	75	50	-
(b)	8.5	2000	-	600	1.	5200	150	200

#1. Data in mg/L except pH, SEM -Standard Error Mean (a) desirable limit (b) permissible limit.

The carbonate and bicarbonate of calcium and Mg^{2+} are producing permanent hardness. The concentration of calcium in groundwater found below the desirable limits, but magnesium was found above the desirable limits given in table 1. The average concentration of sodium ions in groundwater of north zone found beyond the limits of drinking water specification prescribed by WHO. Total 25% of groundwater sample was found beyond the limits for sodium ion in drinking water. The average concentration of fluoride was found above the desirable limit of fluoride in groundwater and east zone found beyond the permissible limits of drinking water.

Fluoride Contamination

The fluoride contents in groundwater of Sareni block are given in table 1. The average concentration of fluoride contents in groundwater samples of centre, north, south, east and west zones of Sareni block were 1.19± 0.19, 1.27±0.28, 1.27±0.28, 1.52±0.19, 1.19±0.17 mg/L respectively. 64% samples of Sareni block were found beyond the desirable limits (1.0 mg/L) of drinking water in Indian standard, while 28% samples were found beyond the permissible limits (1.5 mg/L). The range of fluoride intake between 0.5 to 1.0 mg/L is beneficial for developments and maintenance of healthy teeth, but below that level can cause dental caries and weakness in bone. The natural source is responsible for fluoride contamination in groundwater of India (Subba Rao et al., 2017). According to Kumar and Saxena 2011, fluoride containing minerals (biotite and muscovite) was present in alluvial deposits of clay and mud near the Indo-Gangetic basin. The sediments of the Indo-Ganga bank consist of Old and new alluvium soil.

Newer alluvium is poor and lighter color while old alluvium deposit of clay particle with pale yellow, reddish-brown color and yellowish with kanker deposit with clay layers (Khanna 1992). Figure 2 raveled that fluoride in groundwater was not found uniform in nature, i.e., concentration was ranged from 0.80 to 1.84 mg/L with 60% samples was above the desirable limits (1.0 mg/L) in central zone, 0.64 to 2.21 mg/L with 60% samples was above the desirable limits in north zone, 0.64 to 2.12 mg/L with 60% samples was above the desirable limits in south zone, 1.02 to 2.00 mg/L with 100% samples was above the desirable limits in east zone, 0.94 to 1.88 mg/L with 80% samples was above the desirable limits of fluoride in groundwater of west zone of Sareni block.

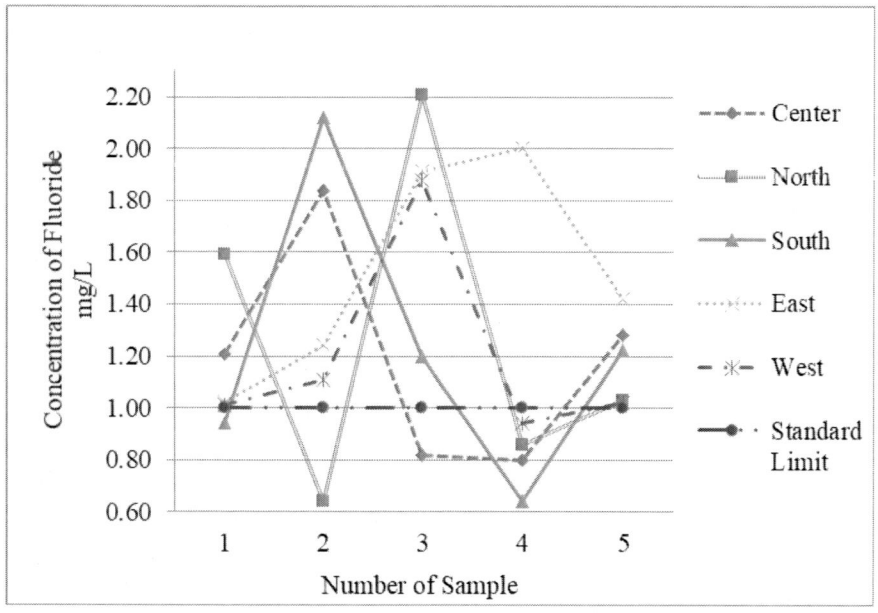

Figure 2. The comparative representation within fluoride level in groundwater of different zone at Sareni block with drinking water standard limit.

The graph of cluster analysis (shown in Figure 9) also distinguish all zone in three deferent clusters based on fluoride contents; cluster one formed with centre and west zone where average fluoride concentration was below the 1.20 mg/L, north and south zone was formed cluster two having average fluoride concentration beyond 1.20 mg/L and east zone differ from both cluster having mean fluoride content above 1.50 mg/L. the special distribution and rate of dissolution of fluoride in groundwater were very from place to place, physical factors such as lithology, slope, climate, soil cover and manmade activity like exploitation of groundwater noticed by (Subba Rao et al., 2017). The soil type of the Ganga basin is dominated with clay particle that can increase the retention time of groundwater and cation exchange capacity both are the major factors for dissolution of fluoride in groundwater.

Statistical Analysis

The correlation analysis is applied to finding the chemical factors that can assist to increase fluoride in ground water are given in Figure 3 to 8. The binary plot plotted between calcium and fluoride in Figure 3 demonstrates no correlation. The correlation graph plotted in Figure 4 and 6 depicts weak positive relation of total hardness and pH with fluoride. The binary plot plotted between bicarbonate alkalinity and fluoride in Figure 5 was shows good positive correlation because both are the result of dissolution of soil mineral. The binary plot plotted between fluoride with TDS given in Figure 8 showing good positive relation because fluoride is also part of dissolved solids. The line graph plotted with fluoride against sodium was given in Figure 7, reveals a positive relation. The concentration of sodium ion was found more than calcium can explaining ion exchange occurs between both elementsand sodium shows higher affinity with fluoride.The Cluster analysis was applied to evaluate the special distribution of fluoride in groundwater based on similarities/dissimilarity, and the dendrogram is rearranged with all the data set low to high concentration of fluoride was given Figure 9. The

Dendrogram of cluster analysis is classifying 5 zones into three clusters. Each cluster is suggesting that geology of zone release equal amount of fluoride in groundwater. The geology of center and south zones both are releasing similar level of fluoride. The geology of north and west zone, both are released similar level of fluoride in groundwater, while east zone was showsmaximum dissimilarities as compare to both above clusters.

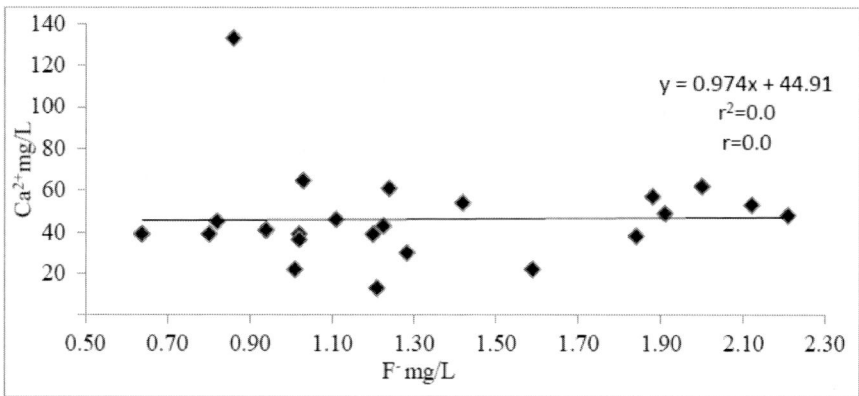

Figure 3. Correlation between fluoride and calcium.

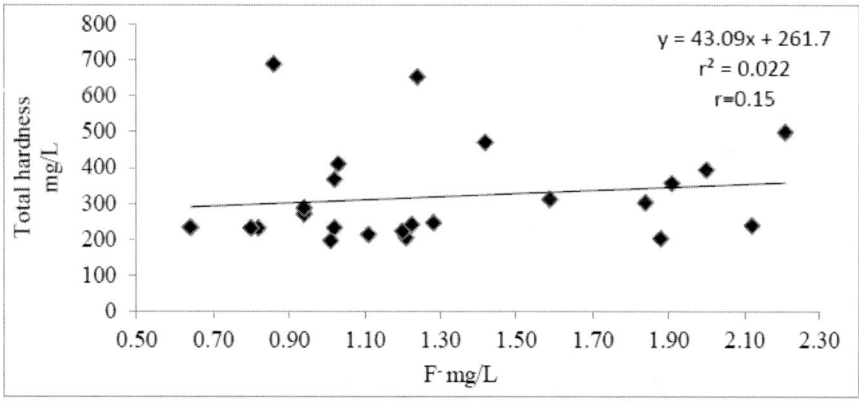

Figure 4. Correlation between fluoride and total hardness.

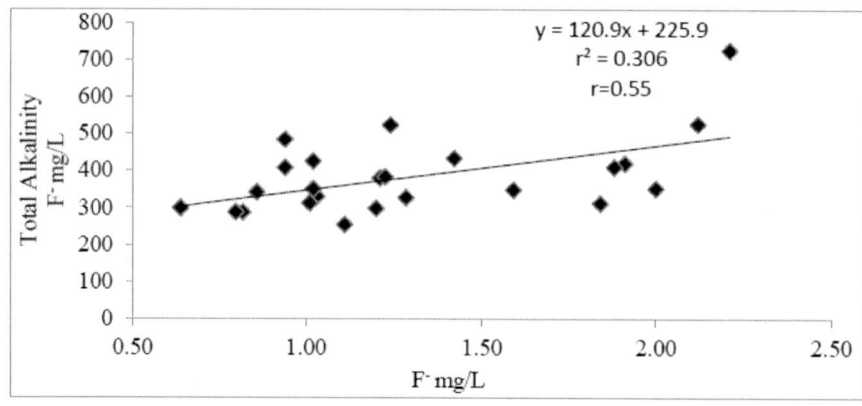

Figure 5. Correlation between fluoride and total alkalinity.

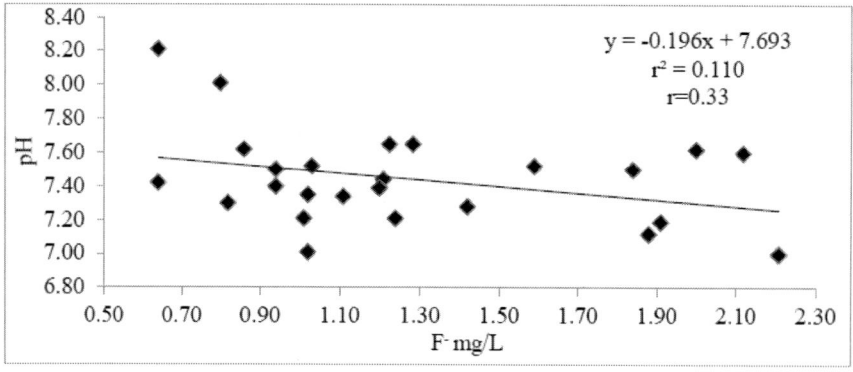

Figure 6. Correlation between fluoride and pH.

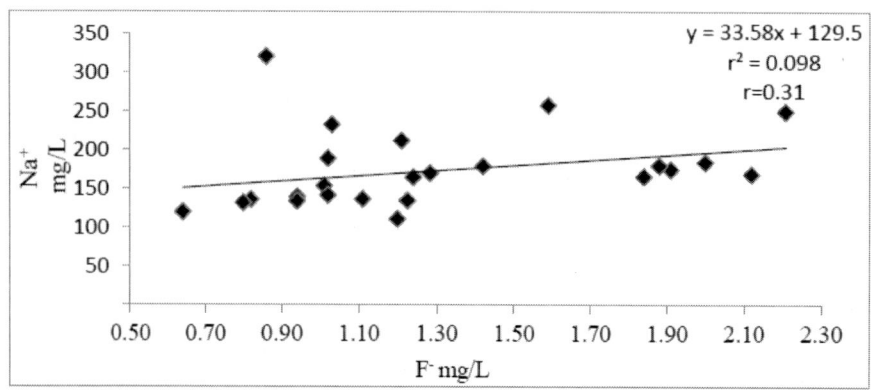

Figure 7. Correlation between fluoride and sodium.

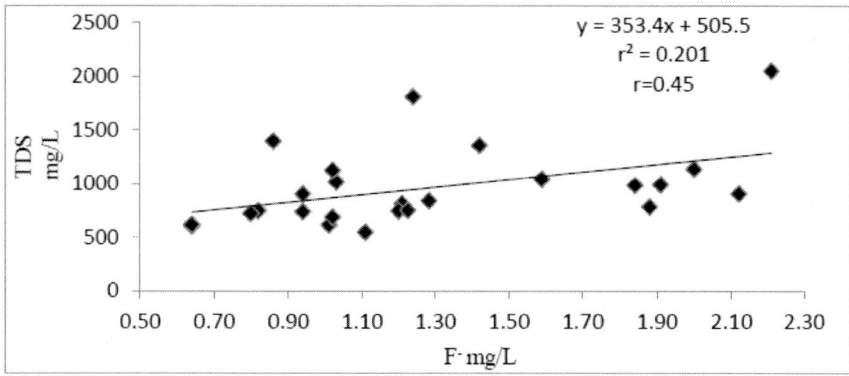

Figure 8. Correlation between fluoride and total dissolved solid.

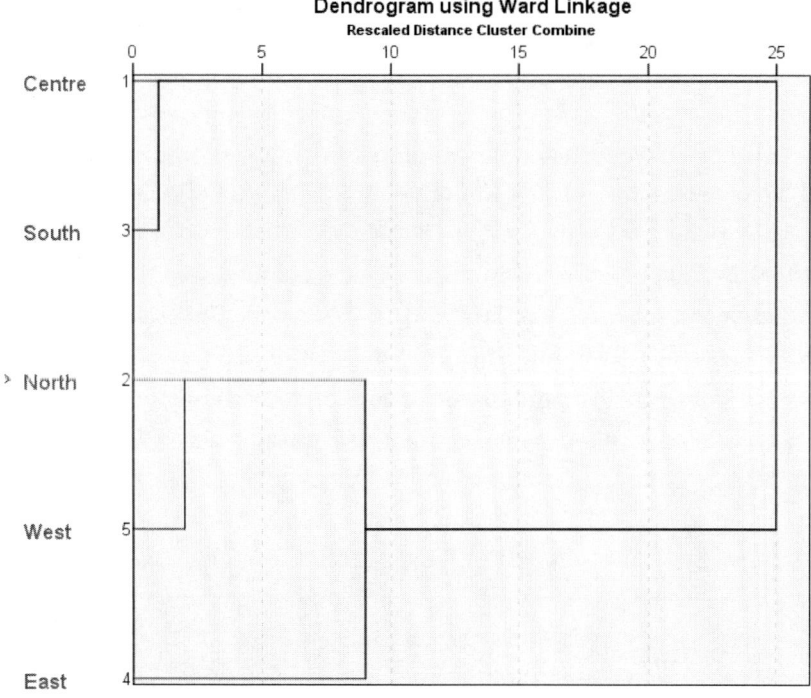

Figure 9. Dendrogram of cluster analysis showing similarity in special distribution of fluoride in groundwater of Sareni block.

Fluoride Exposure Dose Assessment from Drinking Water

This study conducted to assess the risk of fluoride in different age groups of population at Sareni block by computing fluoride exposure dose from drinking water sources is given in table 2. The climate of the Sareni block fall under dry and hot tropical climate and water consumption of the deferent age grouped infants, children and adults were investigated 1, 3, and 6 L per day, respectively. The fluoride exposure dose were calculated for infant of Sareni block ranged from 0.27 to 0.61mg/kg/day for centre zone, 0.22 to 0.74 mg/kg/day for north zone, 0.22 to 0.71 mg/kg/day for south zone, 0.34 to 0.67 mg/kg/day for east zone and 0.31 to 0.63 mg/kg/day for west zone, which was exceeded MRL (minimum risk level) of fluoride (0.05 mg/kg/day) recommended by Agency for Toxic Substances and Disease Registry (Ortiz et al., 1998). Beyond the recommended value evident for dental and skeletal fluorosis. The fluoride exposure dose investigated for children were ranged from 0.12 to 0.28 mg/kg/day for centre zone, 0.10 to 0.33 mg/kg/day for north zone, 0.10 to 0.32 mg/kg/day for south zone, 0.15 to 0.30 mg/kg/day for east zone and 0.14 to 0.28 mg/kg/day for west zone. The exposure dose of fluoride found for adults were ranged from 0.07 to 0.16 mg/kg/day for centre zone, 0.05 to 0.19 mg/kg/day for north zone, 0.05 to 0.18 mg/kg/day for south zone, 0.09 to 0.17 mg/kg/day for east zone and 0.08 to 0.16 mg/kg/day for west zone. The average fluoride intake via drinking water may >8 times higher for infants, ~5 times for children and >2 times higher than recommended MRL value by ATSDR. The results of fluoride exposure dose shown only from a drinking water source, but it may exceed from other sources such as tea, fruit, food, meat, vegetables, and milk, etc. Infants and children of the study area may highly be exposed to elevated level of fluoride intake as compared to adults and fluoride absorption also high because of the development of bones in infants and children. So, the chances of dental and skeletal fluorosis also maximum in infants and children of Sareni block.

Table 2. Evaluation of Fluoride exposure dose through drinking water for inhabitants in different direction of Sareni block

Age groups	Mean Water Consumption (L/day)	Fluoride Level in Groundwater (mg/L)			Fluoride Exposure Dose (mg/kg/day)		
		Min	Max	Mean	Min	Max	Mean
		Center zone					
Infant (6 kg)	1	1.68*	3.68*	2.38*	0.27	0.61	0.40
Children (20 kg)	3	0.8	1.84	1.19	0.12	0.28	0.18
Adults (70 kg)	6	0.8	1.84	1.19	0.07	0.16	0.10
		North zone					
Infant (6 kg)	1	1.28*	4.42*	2.54*	0.22	0.74	0.42
Children (20 kg)	3	0.64	2.21	1.27	0.10	0.33	0.19
Adults (70 kg)	6	0.64	2.21	1.27	0.05	0.19	0.11
		South zone					
Infant (6 kg)	1	1.28*	4.24*	2.46*	0.22	0.71	0.41
Children (20 kg)	3	0.64	2.12	1.23	0.10	0.32	0.18
Adults (70 kg)	6	0.64	2.12	1.23	0.05	0.18	0.11
		East zone					
Infant (6 kg)	1	2.04*	4.00*	3.04*	0.34	0.67	0.51
Children (20 kg)	3	1.02	2.00	1.52	0.15	0.30	0.23
Adults (70 kg)	6	1.02	2.00	1.52	0.09	0.17	0.13
		West zone					
Infant (6 kg)	1	1.88*	3.76*	2.38*	0.31	0.63	0.40
Children (20 kg)	3	0.94	1.88	1.19	0.14	0.28	0.18
Adults (70 kg)	6	0.94	1.88	1.19	0.08	0.16	0.10

#2. (*) We have considered the boiled water was used for preparing liquid milk from milk power as well as drinking purpose for infant children so that we have double the fluoride content due to loss of volume of water.

#3. Minimum Risk Level for fluoride exposure dose is 0.05 mg/kg/day was calculated by ATSDR (Agency for Toxic Substances and Disease Registry).

PREVENTIVE MEASURES

- The fluoride removal techniques such as Nalgonda, ion exchange resin, and membrane filtration technique is suggested for people to the study area before consumption of groundwater.

- High calcium containing food materials are suggested for the inhabitant of high fluoride zone because calcium ion reduces the absorption capacity of fluoride and preventing from dental fluorosis (Dinesh, 1998).
- Consumption of vitamin C containing food items reduces the non-skeletal fluorosis such as neuro-degeneration of brain (Reddy et al., 2018) and prevent from memory deficiency (Raghu et al., 2013).
- The surface water and the rain water contents less fluoride as compared to groundwater, so that surface and rain water may also use after disinfection. Rain water harvesting is the economic and best option for source of drinking water.

Conclusion

The Sareni block of Raebareli district was selected for the assessment of exposure dose of fluoride via drinking water and to investigate the major factor controlling the fluoride bearing groundwater. Raebareli district is situated at the bank of the Ganga River and soils of this area are alluvial. The fluoride contamination in groundwater is increasing gradually thus their associated risk is also increasing vice versa. In groundwater samples, fluoride concentration was found beyond the desirable limits as prescribed by Indian standard for drinking water. Water intake is high in this area due to subtropical and hot climatic condition. The exposure dose of fluoride in inhabitants were observed as 0.40, 0.18 and 0.10 mg/kg/day for infants, children and adults respectively in center zone which is nearly similar in north zone, west zone and south zone i.e., 0.42, 0.19 and 0.11 mg/kg/day, 0.40, 0.18 and 0.10 mg/kg/day and 0.41, 0.18 and 0.11 mg/kg/day respectively but maximum exposure was observed in east zone i.e., 0.51, 0.23 and 0.13 mg/kg/day for infants, children and adults respectively. The Agency for Toxic Substances and Disease Registry prescribed guideline value 0.05 mg/kg/day fluoride from drinking and all other dietary sources.

The exposure dose of fluoride was found nearly 8 times higher for infants, nearly 4 times higher for children and nearly 2 times higher for adults as compared to ATSDR. The inhabitants of the study area were consuming a high dose of fluoride through drinking water, which may at the high risk of dental and skeletal fluorosis. The exposure dose of fluoride was found highest for infants may cause for concern. In dendrogram of cluster analysis, these five zones were grouped into 3 clusters which concluded that geology of center and south zone releases an equal amount of fluoride in groundwater while north and west zone had shown similar level but east zone releases highest level of fluoride in ground water. The concentration of total dissolved solids, alkalinity, total hardness, and magnesium were also beyond the desirable limits; they should be removed by boiling of water. This study also helps to govt. for releasing of a fund under the national program for prevention of fluorosis, which will improve the drinking water quality as well as improving the health of inhabitant of the contaminated areas.

ACKNOWLEDGMENTS

Authors are thankful to Director, CSIR-IITR, Lucknow India for providing all necessary facilities for this work.

REFERENCES

[1] Bhattacharya, H. N., and Chakrabarti, S. (2011). Incidence of fluoride in the groundwater of Purulia district, West Bengal: a geo-environmental appraisal. *Current Science*, 101(2): 152–155.

[2] Chadha, D. K., and Tamta, S. R. (1999). Occurrence and origin of groundwater fluoride in phreatic zone of Unnao district, Uttar Pradesh. *Journal of Applied Geochemistry*, 1: 21–26.

[3] Chakroborty, P.K. (1999). Need of applied research on water quality management. *Indian Journal of Environmental Protection*, 19(8): 595-597.
[4] Dinesh, C. (1998). Fluoride and human health-cause for concern, *Indian Journal of Environment Protection*, 19 (2): 81–89.
[5] Eenadu. (2013). *Relief to fluoride-affected areas in budget 2013 (daily news paper in Telugu)*, p 6.
[6] Gupta, M. K. et al., (1999). Groundwater quality assessment of tehsil Kheragarh, Agra, (India) with special reference to fluoride. *Environment Monitoring and. Assessment*, 59: 275–285.
[7] Jha, S.K., Nayak, A. K. and Sharma Y. K. (2009). Fluoride occurrence and assessment of exposure dose of fluoride in shallow aquifers of Makur Unnao district Uttar Pradesh, India. *Environment Monitoring and Assessment*, 156: 561-566. DOI 10.1007/s10661-008-0505-1.
[8] Karunanidhi, D., Aravinthasamy, P., Subramani, T., Wu, J., and Srinivasamoorthy, K (2019). Potential health risk assessment for fluoride and nitrate contamination in hard rock aquifers of Shanmunanadhi River basin, South India. *Human and Ecological Risk Assessment: An International Journal*, DOI: 10.1080/10807039.2019.1568859.
[9] Khanna, S. P. (1992). Hydrogeology of central Ganga plain, U. P. Gangetic Plain. *Tera Incognita*, pp 23–27.
[10] Kumar, S. and Saxena, A. (2011). Chemical weathering of the indo-gangetic alluvium with special reference to release of fluoride in the groundwater, Unnao District, Uttar Pradesh. *Journal of Geological Society of India*, 77: 459–477.
[11] Kundu, N., Panigrahi, M. K., Tripathy, S., Munshy, S., Powell, M. A. and Hart, B. R. (2011). Geochemical appraisal of fluoride contamination of ground water in the Nayagrah District of Orissa, India. *Environmental Geology*, 41: 451-460. Doi: 10.1007/s002540100414.
[12] Meenakshi and Maheshwari, R. C. (2006). Fluoride in drinking water and its removal. *Journal of Hazardous Materials*, B137: 456-463.

[13] Mondal, N. K., Pal, K. C. and Kabi, S. (2012). Prevalence and severity of dental fluorosis in relation to fluoride in ground water in the villages of Birbhum district, West Bengal, India. *Environment*, 32: 70-84. Doi 10.1007/s10669-011-9374-1.

[14] Muralidharan, D., Nair, A. P., and Satyanarayana, U. (2002). Fluoride in shallow aquifers in Rajgarh Tehsil of Churu District, Rajasthan-An arid environment. *Current Science, 83*: 699–702.

[15] Ortiz, D., Castro, L., Turrubiartes, F., Milan, J. and DaizBarriga, F. (1998). Assessment of the exposure to fluoride from drinking water in Durango, Mexico, using a geographic information system. *Fluoride* 31(4): 183-187.

[16] Raghu, J., Raghuveer, V. C. et al., (2013). The ameliorative effect of ascorbic acid and Ginkgo biloba on learning and memory deficts associated with fluoride exoposure. *Interdicip Toxicology*, 6(4): 217-221.

[17] Raju, N. J., Dey, S., Gossel, W. and Wycisk, P. (2012). Fluoride hazard and assessments of groundwater quality in the semi arid upper Panda River basin, Sonebhadra district Uttar Pradesh, India. *Hydrological Sciences Journal*, 57(71): 1433-1452. DOI: 10.1080/02626667.2012.715748.

[18] Reddy, P.K., Sudhankar, K., and Nageshwar, M. (2018). Vitamin C protects against sodium fluoride induced neurodegeneration in brain of Rats. Intern. *Journal of Advance in Science Engineering and Technology,* 6 (2): 2321-8991.

[19] Sahu, P., Kisku G.C., Singh, P.K., Kumar, V., Kumar, P., and Shukla, N. (2018). Multivariate statistical interpretation on seasonal variations of fluoride-contaminated groundwater quality of Lalganj Tehsil, Raebareli District (UP), India. *Environmental Earth Sciences* 77(484) 1-11. DOI: https://doi.org/10.1007/s12665-018-7658-1.

[20] Sahu, P., Singh, P.K., Kisku G.C., Singh, P and Kumar, V. (2019).An assessment of groundwater quality at Lalganj block, Uttar Pradesh, India: A Water Quality Index approach. *Journal of Emerging Technologies and Innovative Research,* 6 (2), 54-67.

[21] Saxena U, and Saxena S. (2014) Ground water quality evaluation with special reference to Fluoride and Nitrate contamination in Bassi Tehsil of district Jaipur, Rajasthan, India. *International Journal of Environmental Science, 5(1): 144-163.*

[22] Shukla, S. and Saxena, A (2020). Groundwater quality and associated human health risk assessment in parts of Raebareli district, Uttar Pradesh, India, *Groundwater for Sustainable Development*, doi: https://doi.org/10.1016/j.gsd.2020.100366.

[23] Subba Rao, N., Surya Rao, P., Dinakar, A., Nageswara Rao, P. V. and Marghade, D. (2017). Fluoride occurrence in the groundwater in a coastal region of Andhra Pradesh, India. *Applied Water Science*, 7: 1467–1478. DOI 10.1007/s13201-015-0338-3.

[24] Thivya, C., Chidambaram, S., Rao, M. S., Thilagavathi, R., Prasanna, M. V. and Manikandan, S. (2017). Assessment of fluoride contaminations in groundwater of hard rock aquifers in Madurai district, Tamil Nadu (India). *Applied Water Science*, 7: 1011–1023. DOI 10.1007/s13201-015-0312-0.

INDEX

A

acid, 16, 18, 20, 42, 47, 66, 93, 113, 114
adults, x, 64, 108, 113, 122, 124
age, 3, 29, 30, 60, 111, 113, 122
agriculture, viii, 2, 19, 24, 51, 100, 110
alkalinity, 7, 84, 86, 112, 114, 118, 120, 125
allitization, 97
aluminum oxide, 51
anthropogenic activities, 59, 64, 65, 68, 100
anthropogenic source, 17, 19, 93, 98, 100
aquifers, viii, 2, 3, 6, 7, 8, 16, 23, 25, 34, 52, 54, 55, 58, 63, 70, 74, 78, 91, 93, 126, 127, 128
arsenic, v, vii, 1, 2, 3, 5, 6, 7, 8, 9, 10, 11, 12, 13, 14, 15, 16, 17, 18, 19, 20, 21, 22, 23, 24, 25, 26, 27, 28, 30, 31, 32, 33, 34, 35, 36, 37, 38, 39, 40, 41, 42, 43, 44, 45, 46, 47, 48, 49, 50, 51, 52, 55, 71, 72, 74, 75, 76, 78, 108
arsenic poisoning, 6, 8, 44, 48, 49
ascorbic acid, 127
assessment, x, 39, 41, 46, 71, 100, 105, 108, 124, 126, 127

B

atmosphere, 7, 80, 81, 90

Bangladesh, v, viii, 3, 4, 5, 7, 8, 13, 23, 24, 30, 40, 48, 50, 51, 52, 53, 54, 56, 57, 58, 59, 60, 62, 63, 65, 66, 67, 68, 69, 70, 71, 72, 73, 74, 75, 76, 77, 78
behavioral change, 41
bicarbonate, 96, 110, 112, 116, 118
bioaccumulation, 65
biochemical processes, 86, 90
biological activity, 16
biological processes, 15, 34
biological systems, 55
biomass, 23, 100
bisiallitization, x, 80, 97, 100
brick kilns, 67, 72
brick production, 67

C

Ca^{2+}, ix, 21, 64, 80, 84, 87, 88, 92, 93, 94, 95, 98, 99, 112, 115, 116

calcium, 89, 116, 118, 119, 124
Cameroon, v, vii, ix, 79, 80, 81, 82, 83, 84, 91, 97, 98, 99, 100, 101, 102, 103, 104, 105
cancer, 2, 24, 42, 46, 47, 66
capacity building, 38, 40, 45
chemical, vii, ix, 7, 13, 14, 16, 17, 19, 21, 32, 34, 40, 65, 79, 80, 81, 84, 85, 87, 91, 93, 94, 95, 96, 97, 101, 102, 103, 104, 113, 114, 115, 118, 126
chemical characteristics, ix, 80, 81, 87, 101, 102
chemical evolution, 91
chemical properties, 19, 65
chemical structures, 32
chemistry, v, vii, ix, 79, 80, 81, 93, 94, 96, 98, 99, 100, 101, 102, 103, 107
child development, 36
children, x, 5, 36, 55, 56, 64, 108, 111, 113, 122, 123, 124
China, 4, 5, 8, 42, 47, 103, 104, 105, 109
climate, 56, 60, 83, 95, 109, 110, 111, 118, 122
climate change, 56
clinical symptoms, 35
cluster analysis, 112, 118, 119, 121, 125
clusters, 67, 118, 119, 125
community, 7, 32, 40, 43, 45, 48, 49
composition, vii, ix, 16, 29, 79, 80, 81, 91, 93, 97, 98, 103, 104, 114
compounds, 13, 16, 17, 18, 35, 48, 59, 66
consolidation, 32, 74
constituents, 21, 81, 87, 108
construction, 7, 61, 67, 73
consumption, 39, 46, 48, 66, 109, 122, 123
contaminant, vii, 1, 2, 12, 17, 37
contaminated food, 14
contaminated soil, 34
contaminated water, 3, 17, 23, 24, 35, 40, 49, 56
contamination, vii, ix, 1, 2, 5, 8, 13, 15, 17, 19, 24, 30, 32, 33, 34, 37, 48, 50, 51, 52, 53, 55, 56, 65, 68, 71, 74, 76, 108, 110, 116, 124, 126, 128
coordination, 31, 38
correlation, 91, 92, 93, 112, 118
correlation analysis, 118
correlation coefficient, 91, 92
correlation coefficients, 91, 92

D

dendrogram, 118, 125
dental caries, 109, 116
deposition, 90, 98, 111
deposits, 60, 61, 97, 109, 117
depth, 20, 25, 27, 29, 30, 45, 63, 112
detection, 26, 36, 37, 41, 42, 43, 44, 46
detection techniques, 44
developing countries, 14, 31, 68, 101, 114
Dipu Tila aquifer, 54, 58, 59, 61, 62, 63, 64, 65, 69
diseases, 3, 35, 48, 49
disinfection, 90, 124
dissolution, viii, 2, 3, 16, 17, 21, 80, 81, 86, 89, 90, 95, 96, 100, 110, 118
dissolved oxygen, 16
distribution, ix, 6, 20, 50, 58, 59, 80, 85, 94, 118, 121
dominance, 21, 90, 94, 95
donors, 42, 47, 56
drinking water, viii, x, 2, 5, 12, 13, 14, 20, 23, 24, 25, 43, 48, 49, 50, 51, 52, 54, 55, 56, 59, 64, 68, 69, 74, 75, 76, 99, 102, 108, 113, 114, 116, 117, 122, 123, 124, 126, 127

E

economic growth, 59
economic problem, 3

economic status, 42
effluent, 7, 59, 64, 65, 93
eligibility criteria, 59
employment opportunities, 66
environment, vii, ix, 13, 15, 16, 17, 18, 19, 23, 31, 32, 33, 54, 59, 64, 65, 68, 69, 72, 77, 91, 101, 127
environmental conditions, 15
environmental contamination, 65
environmental degradation, 59, 67
environmental impact, v, vii, 53, 54, 63
environmental pollutants, 65
Environmental Protection Agency, 12, 51
evolution, 59, 78, 91, 96, 103
exposure, vii, x, 1, 13, 14, 17, 35, 36, 55, 64, 108, 110, 113, 122, 123, 124, 126, 127
exposure dose, v, x, 107, 108, 110, 113, 122, 123, 124, 126

F

facies, 21, 64, 100, 115
faecal bacteria, 68
fertilizers, 3, 89, 99, 114
fluoride, v, vii, x, 107, 108, 109, 110, 111, 112, 113, 116, 117, 118, 119, 120, 121, 122, 123, 124, 125, 126, 127, 128
fluorosis, v, x, 107, 108, 109, 122, 124, 125, 127
food, viii, 2, 3, 17, 23, 24, 37, 38, 39, 40, 42, 44, 46, 51, 122, 124
food chain, viii, 2, 3, 38, 40, 46
food products, 39, 46
food safety, 38, 42, 44, 46
food security, 37, 51
formation, 3, 16, 21, 29, 38, 86, 89, 93, 96, 97, 110

G

geology, 15, 83, 95, 110, 114, 119, 125
groundwater, v, vii, ix, x, 1, 2, 3, 4, 5, 7, 8, 9, 10, 11, 12, 13, 14, 15, 16, 17, 19, 20, 21, 22, 24, 26, 27, 28, 29, 30, 32, 34, 37, 38, 39, 40, 42, 43, 48, 49, 51, 52, 53, 54, 55, 58, 59, 61, 62, 63, 64, 65, 66, 67, 68, 69, 70, 71, 72, 73, 74, 75, 77, 78,79, 80, 81, 84, 85, 86, 87, 88, 90, 91, 93, 94, 95, 96, 97, 99, 100, 101, 102, 103, 104, 105, 108, 109, 110, 111, 112, 114, 115, 116, 117, 118, 121, 123, 124, 125, 126, 127, 128
groundwater abstraction, 63
groundwater quality, v, vii, ix, 38, 53, 54, 59, 64, 69, 70, 71, 73, 74, 100, 102, 103, 126, 127, 128
growth, 23, 56, 58, 64, 65, 66, 114

H

hardness, ix, 80, 112, 114, 118, 119, 125
harvesting, 37, 56, 124
hazardous wastes, 34
hazards, 2, 38, 49, 109
health, vii, 1, 2, 4, 10, 13, 14, 20, 24, 35, 36, 37, 38, 41, 42, 43, 44, 46, 47, 48, 49, 51, 55, 56, 59, 64, 72, 101, 108, 110, 125, 126
health effects, 14, 24, 51
health problems, vii, 1, 13, 55
health status, 42, 47
heavy metals, 37, 43, 65, 66, 68, 77, 108
hexavalent chromium, 66
Holocene, viii, 2, 3, 60, 61
Holocene Alluvial deposits, 61
human, vii, 1, 3, 8, 13, 14, 24, 35, 47, 48, 55, 59, 64, 65, 66, 70, 80, 108, 113, 114, 126, 128
human development, 47

human development index, 47
human health, vii, 1, 3, 8, 14, 24, 35, 55, 59, 65, 70, 126, 128
hydrochemical facies, 93, 100
hydrofluoric acid, 110
hydrogen, 66, 86
hydrothermal activity, 89

I

industrial effluents, 58, 59
industrial wastes, 7
industrialization, viii, 53
industries, 18, 39, 65, 66, 67, 73, 76
interaction process, 81
ion exchange, x, 15, 16, 80, 92, 96, 100, 118, 123
ions, 16, 81, 87, 88, 89, 91, 92, 96, 98, 110, 116
iron, 16, 18, 20, 30, 34, 43, 45, 56, 66
irrigation, 3, 23, 24, 48, 54, 102
issues, viii, 2, 3, 32, 36, 39, 42, 47, 58, 70, 103

K

K^+, 21, 64, 84, 89, 92, 93, 98, 99
keratosis, 35, 36

L

land subsidence, ix, 54, 63
leaching, 5, 8, 68, 96, 97
leather processing, 65, 72, 77
liver cirrhosis, 65
liver damage, 65
liver disease, 35
livestock, 54

M

magnesium, 89, 112, 116, 125
management, x, 31, 32, 36, 37, 38, 39, 58, 63, 69, 75, 76, 77, 81, 108, 126
manganese, 34, 45, 56, 66
manufacturing, 19, 66, 67, 76, 110
marine component, 98
marine contribution, 99
materials, 7, 8, 32, 34, 64, 67, 81, 110, 124
megacity, v, 53, 54, 70, 71, 74, 75
methylation, 18, 20, 33, 42, 47
micronutrients, 68
microorganisms, 18, 55, 96
migration, 43, 57
mineralization, 86, 95, 109
mitigation, v, viii, 1, 2, 14, 31, 37, 38, 40, 41, 43, 44, 72, 74
mixing, ix, 21, 24, 32, 80
monosiallitization, 97
morbidity, 36
mortality, 55, 114
Mt. Cameroon, v, ix, 79, 80, 81, 82, 83, 84, 97, 98, 100
Mt. Cameroon area, v, ix, 79, 80, 81, 82, 83, 84, 97, 98

N

natural, ix, 2, 3, 5, 13, 15, 17, 18, 19, 26, 34, 41, 50, 54, 55, 57, 63, 64, 66, 67, 78, 80, 89, 90, 93, 95, 98, 100, 102, 108, 110, 114, 116
neurodegeneration, 127
neutral, 14, 15, 86

O

organic compounds, 15
organic matter, 20, 23, 66, 86, 90, 95

over pumping, 63, 69
overexploitation, 54
ownership, 43, 48
oxidation, 13, 15, 16, 18, 19, 21, 24, 34, 90

P

pathogenic microorganisms, 55
pH, 6, 7, 8, 13, 16, 18, 19, 43, 45, 84, 85, 86, 91, 99, 110, 112, 114, 115, 116, 118, 120
physico-chemical data, 85
physico-chemical parameters, 96
physicochemical properties, 103
plants, 2, 7, 15, 18, 23, 32, 33, 35, 49, 59, 68, 76
Pleistocene Madhupur clay formation, 60
pollution, ix, 5, 13, 48, 53, 58, 63, 65, 66, 68, 69, 74, 77, 90, 114
population, vii, viii, 1, 3, 5, 7, 13, 37, 53, 54, 55, 56, 57, 60, 68, 69, 111, 113, 122
population growth, 58, 60
positive correlation, 92, 93, 118
precipitation, 15, 16, 19, 28, 29, 30, 84, 90, 98, 103
public awareness, 37
public health, 14, 22, 46, 59

R

rainfall, 61, 62, 69, 81, 83, 84, 99, 103
rainwater, v, vii, ix, 27, 38, 79, 80, 81, 86, 87, 88, 90, 91, 92, 98, 99, 101, 102, 104
rates of recharge, 62
reactions, 18, 19, 35, 81, 93, 95, 100
recharge, ix, 25, 26, 28, 29, 30, 38, 45, 58, 62, 63, 64, 76, 80, 81, 93, 99, 103
remediation, 15, 24, 34, 69
resource management, viii, 53, 110
resources, 49, 54, 55, 59, 62, 68, 69, 78, 81

risk, v, vii, x, 2, 5, 8, 14, 24, 32, 35, 37, 39, 42, 43, 45, 46, 51, 64, 66, 70, 71, 72, 75, 107, 108, 109, 110, 122, 123, 124, 126, 128
risk assessment, 37, 70, 126, 128
risk management, 45
rock-water interactions, 81
rural areas, 57
rural development, 47
rural population, viii, 2

S

safety, 39, 41, 46, 47
seawater, ix, 80, 98, 99
sediments, viii, 2, 7, 16, 19, 78, 86, 117
sewage, 7, 17, 19, 58, 68, 93, 110
sewage systems, 68
silicate minerals, 16, 17, 19, 89, 95
skin, 10, 24, 35, 36, 109
skin cancer, 24, 35, 36
sludge, 7, 17, 19, 24, 31, 32, 33, 34, 41, 43, 68, 93, 110
soil type, 118
solution, viii, 2, 16, 18, 32, 36, 56, 80, 86, 113
speciation, 13, 16, 18, 64, 78
species, 7, 13, 15, 18, 87, 91, 94, 100
statistical information, 85
sulfate, 45, 66, 99
sulfuric acid, 5, 66
sustainability, 38, 40, 51
sustainable, viii, ix, 2, 14, 24, 31, 32, 37, 43, 54, 69, 70, 71, 74, 75, 76, 78, 128

T

temperature, 16, 61, 83, 110
testing, 37, 38, 41, 44, 45, 49
textile manufacturing processes, 67
textiles, 18, 67

toxicity, 2, 13, 19, 24, 36, 42, 47, 48, 49, 64
transmissivity, 61
treatment, 7, 8, 31, 32, 33, 34, 36, 43, 56, 59
tube wells, 3, 10, 25, 54, 55, 56, 68

U

urban, 7, 19, 56, 102
urban areas, 7, 56
urbanization, viii, 53, 58, 59, 65, 67, 70, 72

V

volcanic areas, 81
volcanic rocks, 83, 89

W

water, vii, viii, ix, x, 1, 2, 5, 7, 8, 13, 14, 16, 17, 18, 19, 20, 21, 23, 24, 26, 27, 28, 30, 31, 32, 33, 36, 37, 38, 39, 40, 41, 43, 44, 45, 46, 47, 49, 50, 51, 53, 54, 56, 58, 59, 62, 63, 64, 65, 66, 67, 68, 69, 70, 72, 75, 76, 77, 78, 80, 81, 84, 85, 86, 87, 88, 89, 90, 91, 92, 93, 94, 95, 96, 97, 98, 99, 100, 101, 102, 103, 104, 105, 108, 110, 112, 113, 114, 118, 122, 123, 124, 126, 127, 128
water chemistry, 93, 96, 98, 101
water demands, 54
water footprint, 66
water policy, 56
water quality, ix, x, 43, 45, 47, 51, 54, 56, 62, 70, 100, 104, 105, 108, 110, 112, 114, 125, 126, 127, 128
water resources, 55, 77, 81
water samples, ix, 20, 21, 27, 28, 68, 80, 84, 85, 86, 87, 88, 90, 91, 92, 93, 96, 97, 98, 99, 113
water supplies, 54, 56, 64
weathering, x, 64, 80, 85, 88, 89, 90, 92, 93, 95, 96, 97, 99, 100, 101, 102, 104, 126
weathering index, 85, 97
wells, 3, 5, 6, 10, 14, 21, 25, 54, 55, 56, 59, 68